THE COMMONWEALTH AND INTERNATIONAL
LIBRARY OF SCIENCE, TECHNOLOGY,
ENGINEERING AND LIBERAL STUDIES

Joint Chairmen of the Honorary Editorial Advisory Board
SIR ROBERT ROBINSON, O.M., F.R.S.
and DEAN ATHELSTAN SPILHAUS

Publisher
ROBERT MAXWELL, M.C.

METALLURGY DIVISION

VOLUME 2

General Editors:
H. M. FINNISTON, D. W. HOPKINS AND W. S. OWEN

LIQUID FUELS

LIQUID FUELS

BY

D. A. WILLIAMS
B.Sc., Ph.D., M.Inst.F., A.F.Inst.Pet.

AND

G. JONES
B.Sc., A.M.Inst.F., A.F.Inst.Pet.

PERGAMON PRESS
OXFORD · LONDON · PARIS · FRANKFURT

THE MACMILLAN COMPANY
NEW YORK

PERGAMON PRESS LTD.
Headington Hill Hall, Oxford
4 & 5 Fitzroy Square, London W.1

THE MACMILLAN COMPANY
60 Fifth Avenue, New York 11, New York

COLLIER-MACMILLAN CANADA, LTD.
132 Water Street South, Galt, Ontario, Canada

GAUTHIER-VILLARS ED.
55 Quai des Grands-Augustins, Paris 6

PERGAMON PRESS G.m.b.H.
Kaiserstrasse 75, Frankfurt am Main

Copyright © 1963
PERGAMON PRESS LTD.

Library of Congress Card No. 62–22046

Set in 10 *on* 12 *pt. Times and Printed in Great Britain by*
ADLARD & SON LTD., DORKING

Contents

Chapter		Page
	FOREWORD BY E. S. SELLERS	vii
	ACKNOWLEDGEMENTS	ix
1	THE MANUFACTURE OF LIQUID FUELS FROM PETROLEUM	1
	Distillation—Fuel Oils—Conversion Processes—Treatment Processes	
2	LIQUID FUELS FROM SOURCES OTHER THAN PETROLEUM	23
	Oil Shales—Coal	
3	PROPERTIES OF LIQUID FUELS	38
	Density and Specific Gravity—Viscosity—Flash Point and Fire Point—Pour Point and Cloud Point—Carbon Residue—Burning Tests—Sulphur Content—Ash Content—Stability and Compatibility—Water Content—Calorific Value	
4	FUELS FOR SPARK IGNITION INTERNAL COMBUSTION ENGINES	50
	Combustion Characteristics—Motor Gasolines—Aviation Gasolines—Vaporizing Oil or Power Kerosine—Alcohols	
5	FUELS FOR COMPRESSION IGNITION ENGINES	65
	Compression Ignition Engines—Characteristics of Diesel Fuels—Specifications for Diesel Fuels—Diesel Fuels from Sources other than Petroleum—Applications of Diesel Fuels	

CONTENTS

6 ATOMIZATION AND COMBUSTION OF FUEL OILS — 76
Oil Burners—Efficient Combustion—Low Temperature Corrosion—Smut Formation and Steel Stack Corrosion

7 THE USE OF LIQUID FUELS IN BOILERS, INDUSTRIAL FURNACES, AND GAS TURBINES — 94
Large Water Tube Boilers—Shell-type Boilers—Industrial Furnaces—Furnace Temperature Control—Furnace Pressure Control—Combustion Control—Open Hearth Furnace—Glass Furnaces—Blast Furnaces—Gas Turbines

8 THE USE OF LIQUID FUELS FOR GAS MANUFACTURE — 118
Characteristics of Town Gas and Industrial Gases—Principles of Oil Gasification—Town Gas Manufacture—Industrial Gas Manufacture—The Use of Liquefied Petroleum Gases

9 OIL FIRED DOMESTIC HEATING APPLIANCES — 144
Oil Burners used in Domestic Heating Appliances—Burner Controls—Oil Burning Appliances used in Domestic Heating Installations—The Storage of Domestic Fuel Oil—Central Heating

10 STORAGE AND HANDLING OF LIQUID FUELS — 162
Storage Tanks—Liquefied Petroleum Gases—Materials having Flash Points below 73°F—Fuel Oils—Tank Heating—Oil Preheaters—Ring Main Systems

INDEX — 175

Foreword

by E. S. Sellers[*]

British Petroleum Company Limited

IN THE teaching of technological subjects in the universities and colleges of technology, the major emphasis must always be on the underlying principles of the subject and the development of design procedures based on these principles. The relatively short time available for formal teaching in our college courses means inevitably that the study of current industrial practices is neglected, and the student must rely on his own reading if he wishes to be informed of these practices. The availability of suitable texts is therefore most important.

This book on Liquid Fuels will be found most useful for the purpose, and it will certainly meet the requirements of those students who are preparing for the Institute of Fuel examinations and for college students of chemical engineering and metallurgy. At the same time the book will be of interest and value to people in industry who use liquid fuels for furnace firing, for internal combustion engines and for gas manufacture. The authors make the subject appear both logical and clear by employing a useful style which combines the description of practice in use, methods of manufacture and the principles which have led to the adoption of fuel specifications.

The use of liquid fuels is expected to increase threefold during the next generation, and the effectiveness of the fuel consuming equipment proceeds at a rate commensurate with the efficiency of combustion or conversion of the fuel used. The improvement of this is attracting some of our most able scientists and engineers, and in presenting much useful information to a wide technical audience, this book will make its own contribution.

[*] Assistant General Manager, Refineries Department. Formerly Professor of Chemical Engineering, University College of Swansea.

Acknowledgements

WE are grateful to Mr. R. B. Southall, the General Manager of BP Refinery (Llandarcy) Limited, for many facilities and encouragement during the preparation of the manuscript, to Mr. T. E. Jenkins for reading the preliminary drafts and proofs, and to those other colleagues at Llandarcy who have given assistance. Our thanks are also due to colleagues at the BP Research Centre and BP Technical Services Branch for their valuable help and suggestions.

We thank the following organizations for permission to reproduce illustrations:

Associated Octel Co. Ltd. (Table 17)
Babcock & Wilson Ltd. (Fig. 23)
British Petroleum Co. Ltd. (Figs. 1, 9, 19, 29, 30)
British Standards Institution (Fig. 52)
Chicago Bridge and Iron Co. (Figs. 49, 50)
David Etchells (Furnaces) Ltd. (Fig. 24)
Iliffe and Sons Ltd. (*Gas Turbines and Jet Propulsion*, Smith, G. G.) (Fig. 28)
Institute of Fuel (Figs. 11, 18, 22, 35, 36, 37, 38, 43)
Iron & Steel Institute (Figs. 26, 27)
Laidlow Drew & Co. (Fig. 16)
Linde Company (Fig. 7)
Peabody Ltd. (Fig. 17)
Perkins (C.M.E.) Ltd. (Fig. 46)
Shell-Mex and BP Ltd. (Figs. 12, 13, 14, 15, 48)
Shell Petroleum Co. (Figs. 19, 51)
Universal Oil Products Co. (Fig. 6)
Wilson (HW) Ltd. (Fig. 46)

D. A. W.
G. J.

CHAPTER 1

The Manufacture of Liquid Fuels from Petroleum

THE MAIN source of liquid fuels is crude petroleum, which occurs naturally in the earth and is essentially a mixture of gaseous, liquid, and solid hydrocarbons. Sulphur, oxygen, and nitrogen derivatives of hydrocarbons may also be present, together with traces of inorganic elements. The proportions of the different hydrocarbon types which are present vary with the source of the crude oil and govern the characteristics of the oil. In extreme cases the oil may range from a light-coloured volatile liquid which, without refining, can be used in internal combustion engines, to a dark asphalt-like material with few volatile components. Natural gas, a mixture of gaseous hydrocarbons, is normally present with the crude oil and accumulates in the upper sections of the oil bearing strata.

Crude oil occurs throughout the world, but the most prolific fields are those of the Middle East. A summary of the world estimated petroleum production and published estimates of proven oil reserves is given in Table 1[1] and the proportion of the world energy demand supplied by oil is shown in Fig. 1.[1]

The majority of crude oils contain varying quantities of gas under pressure and emulsified water. The gas is removed at the wells, and water and inorganic salts are removed by dehydrating and desalting operations prior to the refinery processes.

In a refinery the crude oil is processed to give a series of marketable products. The yields and properties of these products are governed, to some extent, by the nature of the crude oil, but mainly by the processes available for refining and by the demands of the market. As an initial step the wide boiling range crude oil is separated into groups of hydrocarbons of narrower boiling ranges by the process of continuous distillation.

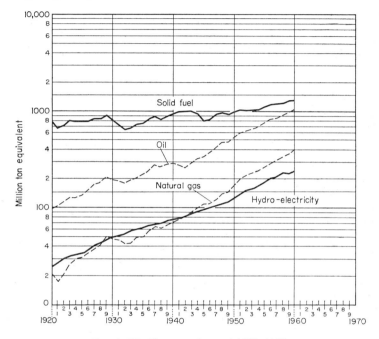

FIG. 1. World energy demand 1920–1960.

Distillation

The hydrocarbons present in the mixture which is crude oil have different boiling points, and by the application of heat it is possible to separate the lighter or more volatile compounds from the heavier ones. Simply described, in the distillation operation[2] crude oil is pumped through a heater to a vertical distillation tower at a temperature which is sufficiently high to vaporize part of the charge. Distillation towers are fitted with internal trays and

TABLE 1
Petroleum Production and Proven Crude Oil Reserves (1960)

	Reserves, million tons	1960 Production million tons
North America	5840	422
South and Central America	3250	180
Western Europe	240	15
Africa	1155	14
Middle East	24,570	256
Far East	1455	27
USSR	4230	146
Eastern Europe and China	270	19
World Total	41,010	1,079

are maintained at fixed high base temperatures and comparatively low top temperatures to give controlled temperature gradients. The vapours thus ascend the tower, bubbling through holes in the trays and contacting liquid flowing down from tray to tray. In this way the vapours become richer in the more volatile components, while the downflowing liquid becomes richer in the heavier components. The fractions of different boiling ranges condense and collect on the appropriate trays at different levels in the tower. The condensation of some or all of the vapours leaving the top of the tower is effected in a water cooled condenser system, and part of this condensate is returned to the tower as "reflux", each tray acting as a heat exchanger between downflowing liquid and ascending vapours. The degree of separation between the different fractions is governed by the number of trays in the distillation column and the amount of reflux returned to the head of the column. The most widely used type of tray in petroleum refining is the bubble-cap type, shown in Fig. 2.

The actual process of crude oil distillation is carried out in a series of heaters and distillation columns, the latter operating at different pressures so that the boiling points of the products are altered to facilitate separation. Thus in high pressure columns it is possible to condense the very light hydrocarbons at temperatures which can be achieved with water at about 80°F, whereas the operation of columns under vacuum enables the higher boiling point materials to be vaporized at temperatures which are sufficiently low to avoid cracking of some of the components.

A modern distillation unit can consist of three stages, a pressure stage, an atmospheric stage, and a vacuum stage. A simplified

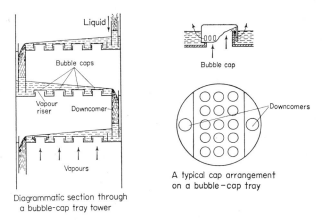

Fig. 2. A bubble-cap tray distillation column.

flow diagram of a typical unit is shown in Fig. 3. The crude oil is initially heated to about 300°F by heat exchange with hot products. It is then passed to the first distillation column, which is maintained at a pressure of about 50 psig in order that the low boiling hydrocarbons such as propane and butane can be condensed in the primary flash distillate. The bottoms offtake from this primary column passes through a pipestill heater, where it is heated to about 600°F before being introduced into the lower half of the secondary column, which is maintained at a pressure slightly above atmospheric. The temperature gradient down this column would be about 240°F to 570°F, and typical offtake temperatures and product boiling ranges are given in Table 2. The maximum operating temperature of an atmospheric column, and thus the amount of distillate and residue obtained, is limited by the temperature at which cracking, or decomposition of the heavier hydrocarbon molecules, occurs. This cracking forms unstable materials which impair the colour and stability of distillates. Further yields of distillates are obtained by distilling

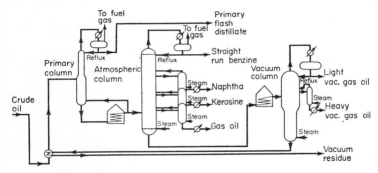

Fig. 3. Flow diagram of a three stage distillation unit.

the atmospheric column residue under vacuum, and this step is used for the production of lubricating oil distillates and feedstock for catalytic cracking. Pressures are normally between 30 and 80 mm mercury absolute, and these are usually obtained by the use of steam ejectors. Steam is injected into the base of both the atmospheric and vacuum towers to reduce the possibility of cracking, and traces of any light components which may be present in the products are removed by steam stripping in small side columns.

The primary flash distillate obtained from the primary distillation stage is further distilled under pressure in a high efficiency "stabilizing" column (not shown in Fig. 3) to give a gasoline containing a controlled amount of butane, and a propane/butane

Table 2
Atmospheric Distillation of a Middle East Crude

Product	Yield on crude % vol.	Boiling range °C	Tray temperature °C	Uses
Straight run benzine	10·0	75–145		Limited quantity blended into motor gasoline and aviation turbine gasoline, reformer feedstock.
Naphtha	4·5	*140–185	130	White spirit, reformer feedstock.
Kerosine	16·5	*180–270	190	Vaporizing oil, burning oil, jet fuels.
Gas oil	6·5	*260–390	280	Diesel fuels, burning oil.
Residue	55·0			Fuel oils, feed to vacuum distillation.

* After stripping in side columns.
The yield of primary flash distillate from the primary column would be about 7% vol.

stream. An additional distillation step may be employed to separate these gases for sale as Liquefied Petroleum Gases. The stabilized gasoline, boiling between 30 and 70°C, may be used in blending low and medium grade motor spirits and aviation turbine gasolines.

The processes which are operated in addition to distillation are those necessary to improve the quality and yield of gasolines, and those necessary for the treatment of vaporizing oils, kerosines, and gas oils. Some refineries include processes for the manufacture of lubricating oil, waxes, and bitumens. A typical flow sheet of a refinery manufacturing fuels from a Middle East crude oil is shown in Fig. 4.

Fuel Oils

Fuel oils cover the range from gas oil to extremely viscous products, and can be generally classified as distillate and residual fuels. The former are blended from atmospheric or vacuum distillates, whereas the latter contain residues or are wholly residues.

Fuel oils are manufactured by blending suitable distillates or distillates plus residues, the blending being carried out in a

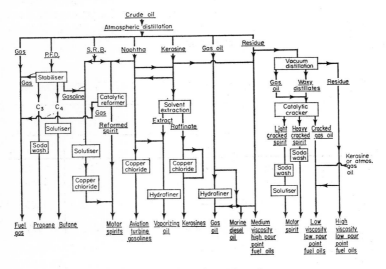

FIG. 4. Typical flow sheet of a refinery.

manner most suitable for each refinery programme. Typical data for fuels meeting the B.S.I. specifications are given in Table 3.[3] In addition to products from distillation units, other components can include gas oils and heavier distillates from cracking processes and extracts from lubricating oil manufacture.

The characteristics of residues are usually determined by distillate requirements, and in refineries where it is necessary to reduce the viscosity of a residue for fuel oil blending, a mild form of thermal cracking, known as "viscosity breaking",[4, 6] is employed. The heavy residue is passed through a cracking furnace where the required conditions are about 900°F and 250 psig, the cracked product being fractionated to yield about 90% of reduced viscosity fuel oil component and a small quantity of gasoline and gas. The process also reduces the pour point of the fuel oil.

It will be seen that, apart from viscosity breaking, the manufacture of fuel oils does not require any special processes, since the fuels are blended from straight distillates, residues, and by-products from other operations.

TABLE 3

Typical Data for Commercial Fuel Oils of Middle East Origin

	B.S.I. Class			
	D	E	F	G
Specific gravity, 60/60°F	0·835	0·935	0·955	0·965
Viscosity, Redwood 1 sec at 100°F	34	200	900	3000
Viscosity, Redwood 1 sec at 122°F	—	122	450	1300
Viscosity, kinematic at 100°F cs	3·3	49	221	737
Viscosity, kinematic at 122°F cs	—	29·5	106	320
Pour point °F	10	20	30	30
Conradson carbon residue % wt	0·05	—	—	—
Flash point (closed) °F	170	200	200	220
Water % vol	Nil	0·05	0·1	0·2
Sediment % wt	Nil	0·01	0·02	0·03
Ash % wt	Nil	0·03	0·05	0·07
Sulphur % wt	Up to 1	Up to 3	Up to 4	Up to 4
Gross calorific value B.T.U./lb	19,600	18,500	18,400	18,300
B.T.U./Imperial gallon	163,700	173,900	176,000	176,000

Conversion Processes

In separation processes, such as distillation, individual hydrocarbons or a series of hydrocarbons of similar physical properties are separated from crude oil or from crude oil fractions. Conversion processes differ from separation processes in that these manufacture new hydrocarbons, or larger amounts of hydrocarbons already present but in small quantities. In this way some surplus fractions from the distillation step can be converted into more desirable products for the market.

The first gasolines used in internal combustion engines were straight-run materials obtained from atmospheric distillation units, and were mixtures of mainly paraffins and naphthenes, with some aromatics and olefins. The amount of gasoline which could be obtained in this way was limited in both quantity and quality (Research Octane Numbers were only of the order of 50), and varied with the source of the crude oil. The expansion of the motor industry resulted in increased demands for gasoline, particularly for increased anti-knock value, and new processes were developed to obtain improved quality by manufacturing gasolines containing more hydrocarbons of the aromatic and iso-paraffinic types. Present day gasolines have Research Octane Numbers in the range 85 to 100.

Both thermal and catalytic cracking are used to break down the heavy hydrocarbon molecules present in the high boiling fractions from crude oil distillations to produce low boiling gasoline fractions which have high anti-knock characteristics. The gases produced in these processes are rich in olefins, and processes such as alkylation[5, 6] and polymerization[5, 6] are utilized to convert these molecules into hydrocarbons boiling in the gasoline range. In addition, thermal and catalytic reforming,[5, 6] and isomerization[5, 6] are used to change the structure of hydrocarbon molecules and thus improve the quality of low boiling fractions such as benzine and naphtha without materially affecting the boiling range. For example, aromatics are formed from naphthenes.

Cracking

In cracking operations the application of heat and pressure to high boiling fractions results in the breakdown of the heavy hydrocarbon molecules. Thus a long chain paraffin will break down into an olefin and a smaller chain paraffin, as shown below.

$$CH_3—CH_2—CH_2—CH_2—CH_2—CH_2—CH_2—CH_2—CH_2—CH_3$$
n-decane
$$\downarrow$$
$$CH_3—CH_2—CH_2—CH_2—CH_2—CH_3 \; + \; CH_2=CHCH_2—CH_3$$
n-hexane   butene

The overall reactions[6] which take place during the cracking operation are far more complex than the above example indicates, since secondary reactions such as polymerization and isomerization also occur, in which the lighter molecules which are initially formed recombine or change structure. In addition there is a complete breakdown of some hydrocarbons to carbon, and thus the overall products of cracking are a range of materials from gas to fuel oil, together with some coke.

The first cracking units were thermal processes, which relied solely on temperature and pressure for the conversion of high boiling fractions. Temperatures of about 900°F and pressures in the range 100–500 psig were used to produce a gasoline of approximately 75 Research Octane Number at a yield of 30%. The use of catalysts resulted in reduced severity of operation and improved product yields and properties. The first commercial catalytic unit was commissioned in 1936 and employed a fixed bed and a number of reactors which were used alternately for the cracking operation and then for regeneration. During the regeneration stage the carbon deposited on the catalyst during the cracking reactions was burned off with air. Processes utilizing moving and fluidized beds were subsequently developed.[5, 6]

A common type of fluidized catalytic cracking process is shown in Fig. 5. In this type of unit a finely divided catalyst is maintained in a fluidized state, so that the properties of the catalyst bed are similar to those of a boiling liquid. The fluidized catalyst can also be transported in a gaseous stream from reactor to regenerator, and vice versa. The particle size of the catalyst is normally in the 5–100 micron range and the material, usually silica/alumina or silica/magnesia, may be of natural or synthetic origin.

Hot catalyst from the regenerator mixes with preheated oil feed, which is thus vaporized and entrains the catalyst into the

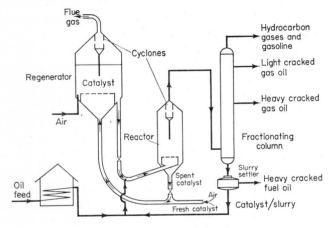

FIG. 5. Fluid catalytic cracker.

reactor. The velocity of the vapours in the reactor is such that a state of fluidization is maintained in which the vapours bubble through the "fluid" catalyst in a manner which resembles a boiling liquid. One of the most important properties of fluidized beds is that a uniform bed temperature is established, temperature variations throughout the reactor normally being less than 5°F.

The coke deposited on the catalyst during the cracking reactions reduces the catalyst activity, so that it is necessary to remove a portion of the catalyst continuously for regeneration. The catalyst passes at a controlled rate from the base of the reactor into a stripping section, where entrained hydrocarbons are removed by steam, and then into the regenerator return line, where it meets a controlled flow of air which entrains the catalyst into the regenerator. Additional air is introduced into the regenerator, where the coke is burned off at a temperature which is not allowed to exceed 1100°F in order to avoid damage to the catalyst.

A certain amount of catalyst is carried with the vapours from the top of the reactor and regenerator beds, and that which does not settle out in the disengaging space above these beds is collected by cyclones and returned. The vapours, containing traces of catalyst, pass from the reactor to a fractionating column and are usually separated into wet gas, unstabilized gasoline, and light and heavy gas oils (cycle oil). For maximum gasoline production the heavy gas oil may be recycled back into the fresh feed and further cracked. In fluid bed units the reactor temperatures and pressures are normally in the range 900–1000°F and 10–15 psig, respectively.

The yield and quality of the gasoline is determined by the feedstock characteristics and by the reaction conditions, the main variables being reactor temperature and pressure, space velocity, ratio of catalyst circulation rate to oil feed rate, and catalyst activity. In a typical catalytic cracking operation the yield would be approximately 48% by volume of gasoline having a Research Octane Number of about 96. Increased yields of gasoline are accompanied by increased yields of gas and coke, and decreased yields of gas oil. The most common feedstocks are distillates in the gas oil range and heavier waxy distillates, the limiting factor being the carbon residue, since too high a carbon residue would result in excessive coke formation in the reactor. The quantity of metal contaminants present is also important because these could result in poisoning of the catalyst and a subsequent decline in activity.

Reforming

In reforming operations the molecular structure of straight run distillates of boiling range 90–200°C is altered to give products with improved octane ratings. In nearly all present day operations the process is catalytic and carried out at elevated temperatures and pressures in the presence of hydrogen. The improvement in octane rating is achieved by converting the naphthenes and straight chain paraffins in the feedstock to aromatics, the main reaction being the de-hydrogenation of the naphthenes. Other reactions include dehydrocyclization (conversion of paraffins to naphthenes and subsequently to aromatics) and isomerization (change in molecular arrangement). Typical reactions are shown below.

(a) *Dehydrogenation*

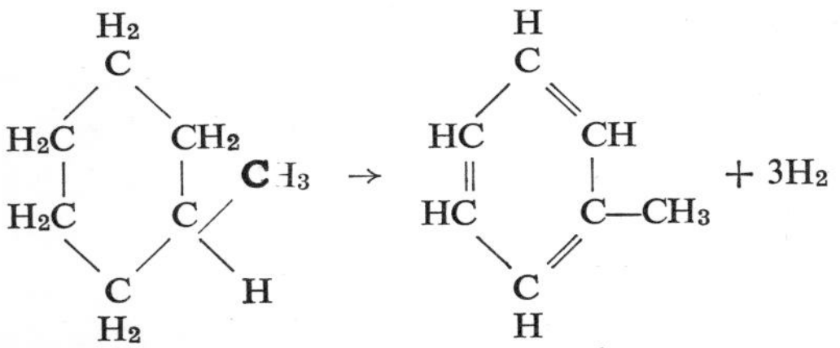

methylcyclohexane → toluene + hydrogen + 3H$_2$

(b) *Dehydrocyclization*

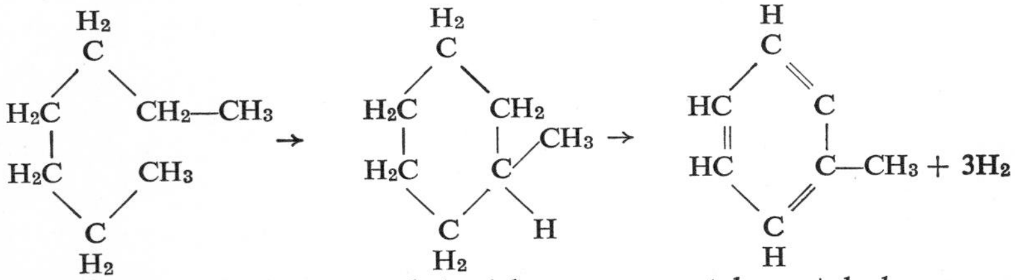

n-heptane → methylcyclohexane → toluene + hydrogen + 3H$_2$

(c) *Isomerization*

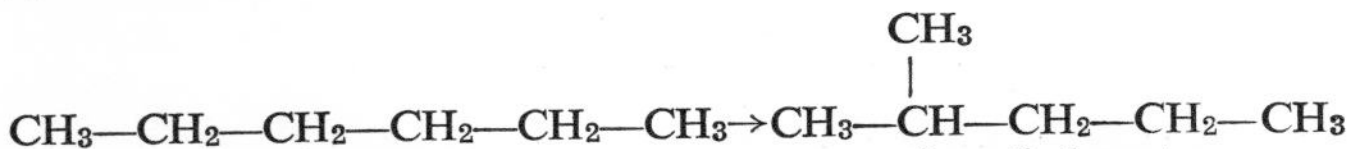

CH$_3$—CH$_2$—CH$_2$—CH$_2$—CH$_2$—CH$_3$ → CH$_3$—CH(CH$_3$)—CH$_2$—CH$_2$—CH$_3$
n-hexane   2-methyl pentane
                                    (iso-hexane)

The improvement in octane rating obtained by the reforming reactions is illustrated in Table 4.

TABLE 4

Hydrocarbon	Research Octane Number
Methylcyclohexane	75
Toluene	120
n-heptane	0
n-hexane	25
2 methyl pentane	73

The hydrogen produced in the above reactions is partially consumed in hydrocracking (cracking and hydrogenation of high molecular weight hydrocarbons) and desulphurization. Some of the excess hydrogen is recycled through the reactors to suppress coke formation and increase catalyst life. The remainder may be used as make-up gas in a catalytic hydrodesulphurization process, or alternatively is blended in the refinery tail gas.

The product, or reformate, is saturated, and the absence of olefins results in a colour stable material which does not form gum. About 90% of the sulphur in the feedstock is removed in the process, and thus the product can be incorporated into gasoline without further refining.

A number of catalytic reforming processes are currently in operation, utilizing fixed or moving catalyst beds. A typical fixed bed process is Platforming, developed by the Universal Oil Products Company, and a flow diagram is shown in Fig. 6. The oil feed is prefractionated to produce a reactor charge of the desired boiling range, this operation also serving to remove water, oxygen and other materials which might poison the catalyst. Catalytic desulphurization of the feedstock may also be employed using excess hydrogen produced in the reforming process.

MANUFACTURE OF LIQUID FUELS FROM PETROLEUM

The treated oil feed is mixed with hydrogen and passed through a series of heaters and reactors containing a platinum-on-alumina catalyst. The reheating furnaces supply the required heat of reaction and thus control the severity of the operation. Operating temperatures are normally in the range 850–1000°F with pressures between 500 and 700 psig, and the main process variables are temperature, pressure, space velocity, and hydrogen recycle rate.

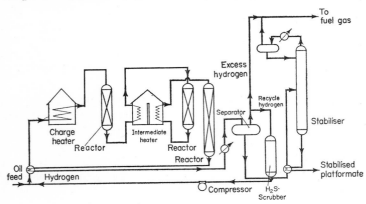

Fig. 6. The Platforming process.

The products of the reaction are separated into gas and liquid streams, the liquid being stabilized to give a reformate of the desired vapour pressure. The gas is compressed and recycled and if hydrodesulphurization of the feedstock is not employed, hydrogen sulphide must be removed.

Reformates consist mainly of paraffins and aromatics, and since most of the paraffins are of low boiling range, a final distillation stage is frequently employed to obtain a concentrate of high boiling, high octane number aromatics. These low volatility materials are then blended with light cracked gasoline fractions to give high octane motor gasolines. Typical reforming operations produce reformates of 85–95 Research Octane Number at a yield of 75–85% wt.

Treatment Processes

The products from the various separation and conversion processes have to be treated to remove any undesirable materials which may be present, for example sulphur compounds from most products, and aromatics from kerosines. These treatments may be carried out on distillates prior to further processing, or on final products in order to meet the necessary specifications. Fuel oils are manufactured by blending suitable heavy components, and although it would frequently be advantageous to carry out a desulphurization treatment, no such commercial process is available at the present time.

Hydrogen sulphide may be present in light hydrocarbon products, and is removed by washing with caustic soda or with weak alkali solutions such as aqueous ethanolamines, as in the Girbotol Process.[5,6] Strong alkali solutions such as caustic soda also remove the lower boiling point mercaptans (up to 100°C), but for the removal of the heavier mercaptans "solutizers" are added to the alkali to increase its solvent power. Some of the solutizing agents used are phenolic materials, salts of fatty acids, and methanol. Typical processes are Solutizer, Unisol, and Mercapsol.[5,6]

A second type of sweetening process, i.e. a process for the removal of the strongly obnoxious mercaptans, oxidizes mercaptans to less offensive disulphides with no reduction in total sulphur content. Some of the most well known employ copper chloride, sodium plumbite (Doctor treatment), sodium and calcium hypochlorite, or lead sulphite as the oxidizing chemicals.

Copper chloride sweetening is one of the processes most widely used and the copper chloride, which acts as a catalyst, is employed as a fixed bed, as a slurry, or in solution. A flow diagram of a slurry type unit is shown in Fig. 7. The oil feed is washed with caustic soda to remove hydrogen sulphide, dried by passage through a tower packed with salt, heated to between 80 and 120°F, mixed with air or oxygen, and passed with the catalyst into a

MANUFACTURE OF LIQUID FUELS FROM PETROLEUM 17

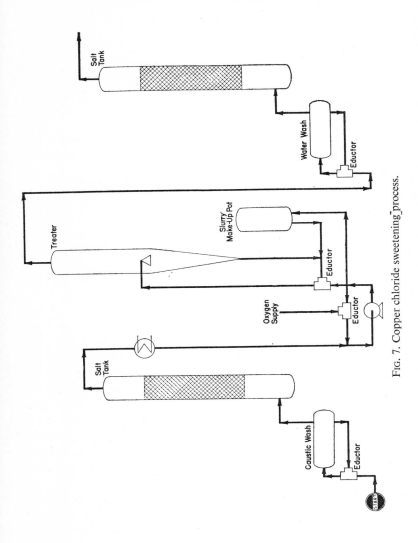

Fig. 7. Copper chloride sweetening process.

treator. The catalyst is recycled continuously and the treated oil is water washed to remove catalyst contamination. The sweetening reactions are

(1) oxidation of mercaptans $\quad 4RSH + 4CuCl_2 = 2R_2S_2 + 4HCl + 4CuCl$

(2) oxidation of cuprous chloride $\quad 4CuCl + 4HCl + O_2 = 4CuCl_2 + 2H_2O$

overall reaction $\quad 4RSH + O_2 = 2R_2S_2 + 2H_2O$

The above processes can be used to sweeten products in the LPG to kerosine range.

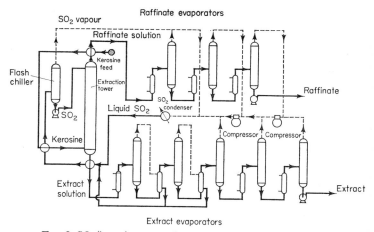

FIG. 8. SO_2/kerosine extraction process. (Edeleanu Process).

Removal of Aromatics

Kerosines have to be treated for the removal of aromatic compounds in order to improve the burning characteristics, and this refining stage is carried out by extraction with liquid sulphur dioxide.[6, 7] Under suitable conditions of temperature and pressure, liquid sulphur dioxide will dissolve aromatic and unsaturated hydrocarbons but will not dissolve paraffins and naphthenes.

Sulphur and nitrogen compounds, which affect odour and colour stability and cause wick deposits, are also removed.

A flow diagram of a typical SO_2/Kerosine plant is shown in Fig. 8. Anhydrous liquid sulphur dioxide is pumped into the top section and dehydrated feedstock into the bottom section of a packed countercurrent extraction tower, where the liquid solvent descends, owing to its higher density, through a rising stream of oil. The temperature is maintained in the range $-10°$ to $15°F$ and the pressure at about 100 psig. Paraffinic raffinate is withdrawn from the top of the tower and aromatic extract from the base, both streams then being passed through heat exchanger and evaporator systems for the recovery of gaseous sulphur dioxide, which is then compressed, condensed to a liquid, and recycled. The raffinate is finally washed with caustic soda to remove the last traces of SO_2. Kerosine is normally treated with 50 to 100% of solvent, and typical results are given in Table 5.

TABLE 5

Kerosine Refining with Liquid Sulphur Dioxide

SO_2 treatment, % vol = 70
Extraction temperature, °F = 10
Extraction pressure, psig = 100

	Feed	Raffinate	Extract
Distillation range, °C	180–250	180–250	180–258
Specific gravity, 60/60°F	0·806	0·785	0·890
Aromatic content, % vol.	18	3	80
Smoke point, mm.	25	40	
Char value, mg/kg	18	2	
Total sulphur, % wt.	0·25	0·02	0·85
Yield, % vol.	100	80	20

The high aromatic extract is a useful component of tractor vaporizing oil owing to the high anti-knock value of the aromatics. In some cases the extract is refined for the manufacture of special solvents.

Hydrogen Processing

Hydrogen treating processes, as distinct from hydrocracking and hydrogen reforming, employ mild hydrogenation conditions to remove sulphur, non-hydrocarbons (oxygen, nitrogen, halogens), and unsaturated compounds. In addition to sulphur reduction, other properties which are improved by hydrogenation are odour, colour, char value of kerosine, and carbon residue. In most cases the hydrogen required for the process is obtained as a by-product from reforming processes. Hydrogen is used to treat

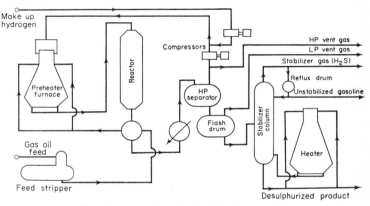

FIG. 9. Process flow in the hydrofiner unit.

distillates such as reformer feedstocks, vaporizing and heating oils, diesel fuels, and catalytic cracker feedstocks.

A large number of processes exist, in all of which the oil is mixed with recycle and/or fresh hydrogen, heated, and passed through a reactor containing a catalyst which requires periodical regeneration. Typical processes[5, 6] are Hydrofining, Hydrodesulphurization, Trickle Hydrodesulphurization, Ultrafining, and Autofining. The Autofining process differs from the others in that the hydrogen required is generated by dehydrogenation of part of the naphthenes in the feedstock.

A flow diagram of the Hydrofining Process,[8] which was developed specifically for the treatment of gas oils, is given in Fig. 9. The feed is first stripped to remove any dissolved air which might result in gum formation, and then passed through a heater where it is mixed with recycle and fresh hydrogen. The mixed stream enters the fixed bed reactor, which contains activated cobalt and molybdenum oxides on alumina catalysts at a temperature of 760–780°F, the pressure in the reactor being 1000 psig. The product is then cooled and passes to a separator from which hydrogen-rich gas is recycled by compressors. The liquid product enters a flash drum before being stripped of hydrogen sulphide in a stabilizing column, the additional heat required for this operation being supplied from a fired heater. Results are given in Table 6.

TABLE 6

Hydrofining Gas Oil

	Feedstock	Product
Specific gravity, 60/60°F	0·842	0·830
Sulphur content, % wt.	0·97	0·08
Yield, % vol.	—	99·0

Desulphurization treatment of gas oils is most desirable owing to the corrosive nature of sulphur compounds in diesel engines and also to the troubles which might be experienced with sulphur products in combustion appliances.

The Gulf HDS Process[6,9] has been developed for the hydrodesulphurization of residual stocks, but neither this nor any other process has yet been shown to be commercially attractive for this treatment.

References

1. *Statistical Review of the World Oil Industry*, The British Petroleum Company Limited, London 1960
2. W. L. NELSON, *Petroleum Refinery Engineering*, McGraw-Hill (1958)

3. J. S. WILDING and E. BRETT DAVIES, *Conf. on Major Developments in Liquid Fuel Firing* 1948–1959, Inst. Fuel (1959)
4. M. G. BOONE and D. F. FERGUSON, *Oil and Gas Journal*, **52** (46), 166 (1954)
5. Process Handbook, *Petroleum Refiner*, **31**, No. 9 (1960)
6. *Modern Petroleum Technology*. The Institute of Petroleum, London (1962)
7. V. A. KALICHEVSKY and K. A. KOBE, *Petroleum Refining with Chemicals* Elsevier Publishing Company, Amsterdam (1956)
8. D. A. SUTHERLAND and F. W. WHEATLEY, *Petroleum Engineer*, C.37, March (1956)
9. *Petroleum Refiner*, **39**, 4, 143 (1960)

CHAPTER 2

Liquid Fuels from Sources Other than Petroleum

ALTHOUGH petroleum is in general the cheapest and largest source of liquid fuels, oil shales and coal are being processed in some countries to produce substitute fuels.

Oil Shales

Deposits of oil shales exist extensively throughout the world and the reserves of recoverable oil are greater than the known reserves of crude oil. The largest deposits are in the U.S.A., and those in Colorado, Utah and Wyoming are estimated as equivalent to 500,000 million barrels of oil. Extensive deposits also exist in France, Estonia, Sweden, Tasmania, Australia, Manchuria and South Africa, but these vary in size, oil content and oil composition. In Scotland, in the Lothians, shale reserves are estimated at between 500 and 900 million tons.[1]

The true oil shales contain no free oil but during destructive distillation a crude oil similar to crude petroleum is produced, together with gases which include ammonia. The oil is formed from a complex mixture of organic compounds termed "kerogen", and varies in quantity from traces to 150 gallons/ton of shale, this latter amount having been obtained from deposits in New South Wales. Shales are being worked in the countries listed above, but the recovery of oil from shale cannot compete economically with petroleum refining. In Sweden, for example, both regular and premium grade gasolines are produced from shale, and the production in one year has included 250,000 bbl of finished gasoline, 500,000 bbl of domestic and industrial fuels, and 6 million gallons of C_3/C_4 hydrocarbons.[2] In most cases shales are obtained by conventional mining methods but in Sweden experiments have been made in the electrical heating of shale underground, thus distilling the shale *in situ* to eliminate the expensive mining procedure.

In Scotland the shale oil industry was developed in the second half of the nineteenth century and has continued until today, when about 1·25 million tons/annum of shale are processed. Shale is dark brown to black in colour, has a laminated structure and contains approximately 80% ash and 1·5% nitrogen. The distillation process in continuous vertical retorts is a typical one and crushed shale is heated at retort temperatures which vary from about 900°F in the top to about 1300°F in the base. Heat is supplied externally by gases obtained from the distillation, and, internally by the injection of a steam/air mixture whereby part of the residual carbon in the shale is burned. In addition the hydrogen formed in the water gas reaction reacts with residual nitrogen to form ammonia.

Average yields of products, per ton of shale, are 2000 ft^3 of permanent gas, 90 gallons of ammonia liquor and 25 gallons of crude oil. The crude oil is dark green in colour, has a setting point of about 90°F and a specific gravity of 0·860–0·890. It differs from the average crude petroleum in that it contains a greater proportion of unsaturated compounds, nitrogen bases and phenolic compounds.

The refining of crude shale is similar to that of crude petroleum and the properties of the main products manufactured from Scottish shale are shown in Table 7.*

Coal

The liquid fuels can be obtained from coal by three different processes:

* Production was discontinued in June 1962.

LIQUID FUELS—SOURCES OTHER THAN PETROLEUM

TABLE 7

Typical Yields and Properties of Shale Fuels

	Motor spirit	Diesel oil
Yield—gal/ton of shale	3·0	11·0
Specific gravity at 60/60°F	0·72	0·84
Octane number, motor method—clear	60	—
Cetane number	—	53
Total sulphur % wt.	0·10	0·30

(1) Carbonization. (2) Hydrogenation. (3) Gasification to form synthesis gas, followed by hydro-carbon synthesis.

In carbonization processes coal tar and crude benzole are obtained as by-products during the manufacture of coal gas and coke. The total quantity of these liquid products is about 8% wt of the coal carbonized, the low yield being the direct result of the low percentage hydrogen present in coal as compared with crude petroleum. The ratio of carbon to "effective hydrogen" for various fuels is shown in Table 8, "effective hydrogen" being defined as the hydrogen available above that required for combination with the oxygen, nitrogen and sulphur in the fuel.[3]

TABLE 8

Ratio of Carbon to Effective Hydrogen in Various Fuels

Bituminous coal	22 : 1	Heavy fuel oil	7·5 : 1
Coal tar	16·5 : 1	Motor spirit	6 : 1
Creosote	13 : 1	Butane	4 : 1
Benzole	12 : 1	Town gas	3 : 1

Increased yields of oil products can be obtained from coal, therefore, by the addition of hydrogen, and this is achieved (a) by destructive distillation in the presence of hydrogen and (b) by complete gasification in the presence of steam to form carbon monoxide and hydrogen, i.e. synthesis gas, which is then converted to liquid products via such processes as the Fischer-Tropsch.

Coal Tar

Coal tar is obtained mainly from high temperature gas retorts and coke ovens during the manufacture of town gas and coke, a further source of supply being from low temperature carbonization processes, but in Britain only about 1·5% of the total tar made is obtained in this way.

The yield and properties of a coal tar depend on the temperature and condition of carbonization. High temperature tars are viscous, of relatively high ash content, and contain large quantities of naphthalene and free carbon, whereas low temperature tars are fairly fluid and contain only small amounts of these latter materials. The yield of tar increases with decreasing carbonization temperature, a maximum of 18–20 gallons per ton being obtained in the low temperature range 550–600°C. The yields of tars from other sources are as follows:

High temperature carbonization, gas works, vertical retorts 12–14 gal/ton.

High temperature carbonization, gas works, horizontal retorts 9–11 gal/ton.

High temperature carbonization, coke ovens, about 8·5 gal/ton.

In Britain about 3 million tons of crude tar are produced per annum, approximately 8% being used directly as a fuel, the remainder being distilled to manufacture a series of fuel oils of similar viscosity to those obtained in the petroleum industry. About 1 million tons per year of different grades of oils are obtained in this way as compared with about 20 million tons of similar petroleum fuels. A scheme of tar distillation is shown in Fig. 10.[4]

LIQUID FUELS—SOURCES OTHER THAN PETROLEUM

Various processes are used for distilling coal tars[4] and these include heating in a series of pot stills maintained at different temperatures and fitted with vapour arms and condensing equipment, and in pipestills with the vapour passing to conventional fractionating equipment. Although Fig. 10 shows the complete range of normal distillates, the extent of any tar distillation will depend on the market requirements. The fuels are designated C.T.F. 50, 100, 200, 250, 300 and 400, the nomenclature giving an

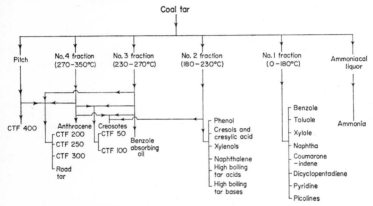

FIG. 10. Products obtained by the distillation of coal tar.

indication of viscosity, and the number its recommended atomizing temperature in degrees Fahrenheit. Thus for C.T.F. 50 this temperature is below the normal ambient temperature, whereas C.T.F. 200 requires heating to 200°F for satisfactory atomization, i.e. a maximum viscosity of 100 Redwood seconds. C.T.F. 50 and C.T.F. 100 are blends of medium and heavy oil distillates and contain no undistilled tar. The end product of the distillation, C.T.F. 400, is a medium soft pitch, and the intermediate fuels are produced by blending or oiling back this latter material with the heavier tar distillates. Specifications for the principal coal tar fuels are given in Table 9 and typical characteristics of these fuels in Table 10.[5]

TABLE 9

Specifications for Liquid Coal Tar Fuels

	C.T.F. 50	C.T.F. 100	C.T.F. 200	C.T.F. 250
Viscosity—Redwood 1 sec	60 max at 100°F	100 max at 100°F	1500 max at 100°F 1000 min at 100°F 100 max at 200°F	1500 max at 160°F 100 max at 250°F
Liquidity	Completely liquid at 32°F	Completely liquid at 90°F		
Flash point (closed) °F	150 min	150 min	150 min	150 min
Gross cal. value B.T.U./lb	16,500 min	16,500 min	16,250 min	16,000 min
Ash content % wt.	0·05 max	0·05 max	0·25 max	0·3 max
Sulphur content % wt.	Of the order of 1	Of the order of 1	Of the order of 1	Of the order of 1

TABLE 10

Typical Characteristics of Coal Tar Fuels

	C.T.F. 50	C.T.F. 100	C.T.F. 200	C.T.F. 250
Cal. value B.T.U./lb, gross	17,050	16,850	16,520	16,410
Therms per ton	382	377	370	367
Sp. gravity at 60/60°F	1·02	1·04	1·14	1·17
Flash point, °F	180–200	180–200	200–220	200–220
Specific heat	From 0·34 to 0·40	From 0·34 to 0·40	From 0·34 to 0·40	From 0·34 to 0·40
Sulphur % wt.	0·75	0·77	0·82	0·84

Coal tar fuels may be used in ordinary industrial furnaces and boilers, but their most important use is in direct fired metallurgical furnaces, where their low sulphur content and high flame emissivity make them excellent fuels.[4] They consist mainly of aromatic hydrocarbons and have higher carbon/hydrogen ratios than the mainly paraffinic petroleum fuels. Their calorific value, on a weight basis, is lower than the corresponding petroleum fuels.

Benzole

The crude spirit recovered from both coal gas and coal tar is termed crude benzole. Originally the product was obtained almost entirely from the light oil or first fraction recovered during the distillation of coal tar. Today the biggest percentage is produced by scrubbing the coal gas with a high boiling point wash oil, such as gas oil, or by adsorption in activated carbon. The relative quantities of benzole produced from these sources is shown in Table 11.

TABLE 11

Production of Crude Benzole in Great Britain

Source of crude benzole	Gal $\times 10^{-6}$
Gas stripping—at coke ovens	83·5
—at Gas Works	24·4
Tar distillation	12·0
Total (1958)	119·9

The yield of crude benzole varies with the type of coal used and the carbonization conditions but averages about 3 gallons/ton of coal. It is water white in colour and is classified by the percentage distilling below 100°C, e.g. 90's benzole. Benzoles consist almost entirely of benzene, toluene, and xylenes, with smaller amounts of paraffins and naphthenes and also some sulphur compounds. The amount of aromatics varies with the carbonization process as shown in Table 12.

Crude benzole is refined by initially washing with sodium hydroxide solution to remove tar acids, and then with sulphuric

TABLE 12

Aromatic Content of Different Benzoles

Carbonization process	Approximate % vol.
Coke oven retorts	90
Gas industry—Horizontal retorts	90
Gas industry— Vertical retorts	70
Low temperature carbonization	30

acid to remove nitrogen bases, unsaturated hydrocarbons, and some sulphur compounds. The product is then redistilled to remove the remaining sulphur compounds, mainly carbon disulphide, and to obtain a distillate boiling in the range 80–150°C. This material may then be incorporated into motor spirit (The National Benzole Association's specification for motor benzole is shown in Table 13), or further distilled to separate the individual hydrocarbons.

TABLE 13

National Benzole Association's Specification for Motor Benzole

Specific gravity at 60/60°F	0·870–0·886
Distillation °C	
Not less than 60% vol. at	100
Not less than 95% vol. at	155
Crystallizing point, °C, not above	5
Sulphur, % wt not more than	0·4

Benzole can be used satisfactorily in internal combustion engines but is normally blended with petroleum gasolines since when used alone it has the following disadvantages:

(1) its initial boiling point is too high to permit ease of starting an engine;

LIQUID FUELS—SOURCES OTHER THAN PETROLEUM

(2) its freezing point is too high, about $-5°C$;
(3) because of its high aromatic content it is difficult to burn without carbon deposition.

The octane number of benzole is about 115 (Research Method), and it is a valuable component for upgrading low octane number gasolines.

Hydrogenation of Coal

The destructive hydrogenation of coal during carbonization can increase the yield of oil products from about 8% to 75% weight of the coal carbonized. This operation was first carried out successfully by Bergius in 1914, and the products obtained included a small quantity of gasoline with some heavier oils, phenols, and pitch.[6]

The first commercial unit was erected in Germany in 1927, and was designed to produce 100,000 tons of gasoline per annum from brown coals. During World War II, hydrogenation plants provided the majority of Germany's liquid fuel requirements and by 1944 3·5 million tons/annum were being produced from coal, coal tar and heavy petroleum residues.

The first commercial unit for hydrogenating bituminous coals was erected in the United Kingdom by the Imperial Chemical Industries Ltd. in 1935. The process operated in two stages,[7]

(1) Liquid phase hydrogenation to give a small amount of petrol and "middle oil".
(2) Vapour phase hydrogenation of the "middle oil" to petrol.

In the first stage the coal was ground and mixed with a heavy recycle oil and catalyst, and then passed with hydrogen at about 250 atmospheres through a preheater and into a convertor which was maintained at about 475°C. The exothermic reactions which occurred were controlled by injecting cold hydrogen into the convertor. The reaction products were cooled and separated from the unreacted coal, the residual hydrogen recycled, and the oil products separated into stabilized petrol, middle oil and heavy oil. The heavy oil was recycled and the middle oil further hydrogenated in two vapour phase convertors. These were fixed bed units operated at 250 atmospheres with a hydrofining stage and a hydrocracking stage, the temperatures being maintained at about 420°C and 380°C respectively. The product was soda washed before the hydrocracking stage to remove phenols and nitrogen bases, and the final product was fractionated to give a stabilized petrol and heavier fractions which were recycled.

The yield of regular grade motor spirit from a suitable clean coal of about 84% carbon content and 2·5% ash content was 46·1%, i.e. one ton of petrol required 2·17 tons of clean coal. If the coal required for hydrogen production is taken into account, a total of 4·5 tons of coal would be required for 1 ton of petrol.[7]

The I.C.I. Unit could not produce liquid fuels as economically as those obtained from crude petroleums, and in 1939 the unit was converted to the hydrogenation of creosote tar,[6] this material being easier to hydrogenate since it is a liquid and has a lower carbon/hydrogen ratio than bituminous coal. Until 1954 between 35 and 45 million gallons of aviation and motor spirits were produced annually, but subsequently production declined because of the difficulty in competing economically with the production of petroleum oils. The plant was finally shut down in 1958. Today the majority of the coal hydrogenation plants in Europe are being used for the hydrogenation of petroleum products, e.g. the hydrocracking of residues.

Gasification of Coal and the Fischer-Tropsch Synthesis

Liquid fuels can also be produced from coal by first gasifying the coal to produce carbon monoxide and hydrogen, followed by the catalytic conversion of these gases by the Fischer-Tropsch Synthesis. In this operation the synthesis gas is passed over an alkali-promoted iron oxide catalyst at temperatures in the range 200–350°C and at pressures of up to 750 psig. The products are a mixture of liquid and gaseous paraffin and olefin hydrocarbons, together with some oxygenated compounds such as alcohols. The composition of the products varies with the conditions employed and with the composition of both the catalyst and the synthesis gas. All the reactions which occur are exothermic and may be represented as follows:

Synthesis of paraffins
$$n\, CO + (2n + 1)\, H_2 \rightarrow C_n H_{2n+2} + nH_2O$$
Synthesis of alcohols
$$n\, CO + 2n\, H_2 \rightarrow C_n H_{2n+1} OH + (n - 1)\, H_2O$$
Synthesis of olefins
$$n\, CO + 2n\, H_2 \rightarrow C_n H_{2n} + nH_2O$$

The main impurities in synthesis gas are hydrogen sulphide, traces of organic sulphides and carbon dioxide, and these are removed prior to the conversion process. A typical method is washing with mono or diethanolamine.

The Fischer-Tropsch process as a source of fuels is not a success economically because of the cheaper manufacturing route of similar fuels from crude petroleum. Should the reserves of petroleum become depleted, however, this process could become one of the main sources of liquid fuels and considerable research work has been carried out in this field together with some commercial development.[8] At present the major portion of the operational costs, about 80%, is taken up by the manufacture and purification of the synthesis gas. One of the best methods of coal gasification for this purpose is the Lurgi process which operates at 20 to 30 atmospheres and thus provides a gas at the required pressure for synthesis.

The world's largest plant for the manufacture of liquid fuels and chemicals from coal is sited near Johannesburg and operated by the South Africa Coal, Oil and Gas Corporation (SASOL). Because of the plentiful supply of cheap coal and the lack of crude oil reserves in South Africa this plant would appear to be capable of working economically. The unit has been on stream for six years and supplies about 20% of South Africa's total petroleum demands.

Sub-bituminous coal is gasified with superheated steam and oxygen in 9 Lurgi gasifiers operating at about 380 psig. These handle 3050 tons of coal per day and produce 162 million SCF of raw synthesis gas.[9] Dust, tar and liquid hydrocarbons are also produced, and these are separated from the gas prior to gas purification for the removal of H_2S and CO_2. This latter operation is carried out using the Rectisol process, in which the gas is washed with methanol at an elevated pressure and about $-70°F$. The purified gas, containing 56% hydrogen, 27% carbon monoxide, 14% methane and small quantities of carbon dioxide and nitrogen, is then passed either through an Arge fixed bed synthesis unit or is reformed to convert the methane present to carbon monoxide and hydrogen before processing in a Kellogg fluidized bed synthesis unit. The reforming operation is carried out using steam and oxygen in the presence of a nickel catalyst at temperatures and pressures in the ranges 1600–2200°F and 300–350 psig respectively. The ratio of hydrogen to carbon monoxide is controlled at the required level by maintaining the necessary operating conditions.

The main operating variables in the Fischer-Tropsch synthesis are reactor temperature and pressure, recycle ratio, i.e. ratio of

TABLE 14

Operating Conditions of Fischer-Tropsch Reactors

Unit	Hourly space velocity	Temperature °C	Pressure atm	Catalyst
Arge fixed bed	500	Up to 270	25–30	Precipitated iron.
Kellogg fluidized bed	400*	300–340	25–30	Millscale or similar iron oxide.

* Includes disengaging space—about double this value on catalyst space.

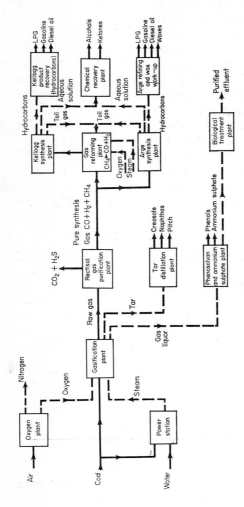

Simplified flow diagram of Sasol works.

product gases recycled to gas feed, composition of the feed gas, and type of catalyst. Typical operating conditions of the Kellogg and Arge synthesis units are given in Table 14.[9]

Operation at the higher temperatures, as in the Kellogg unit gives increased yields of high octane gasoline and comparatively low yields of heavier oils and waxes. In the Arge unit, which is operated at lower temperatures, there is a smaller yield of a poorer quality gasoline but bigger yields of the heavier products.

The exothermic reactions are controlled in both synthesis units and the respective products are separated into gas, water and oil phases. The hydrocarbons are separated in petroleum refinery type units[10] and an example of the products obtained from a Kellogg synthesis plant is shown in Table 15. A simplified flow diagram of the SASOL works is shown in Fig. 11.[11]

TABLE 15

Products Recovered from Kellogg Synthesis Unit

Product	Barrels/day
L.P.G.	24
Gasoline (Research Octane Number, clear 86–90)	3260
Diesel oil	142
Waxy oil	46
Methanol	14·5
Ethanol (crude)	12
Ethanol (pure anhydrous)	302
Methyl ethyl ketone	21·8
Acetone (pure)	15·8

References

1. S. E. COOMBER, Oil Shales, *Handbook of the Petroleum Industry*, George Newnes, London 1958, p. 114
2. *Petroleum Week*, 7, August 8, 1958
3. *Report of the Committee on Coal Derivatives*, Her Majesty's Stationery Office, August 1960, p. 9

4. *Coal Tar Fuels*, Association of Tar Distillers, 1960
5. J. S. WILDING and E. BRETT DAVIS, *Conf. on Major Developments in Liquid Fuel Firing* 1948–59. Inst. Fuel 1959
6. J. S. S. BRAME and J. G. KING, *Fuel*, Edward Arnold, London 1956, p. 272
7. *Report of the Committee on Coal Derivatives*, Her Majesty's Stationery Office, August 1960, p. 69
8. J. H. FIELD, H. E. BENSON and R. B. ANDERSON, Synthetic liquid fuels by the Fischer-Tropsch process, *Chemical Engineering Progress*, **56,** 4, 44, 1960
9. L. W. GARRETT Jr., Gasoline from coal via the Synthol Process, *Chemical Engineering Progress* **56,** 4, 34, 1960
10. R. A. LABINE, Taking a look at coal chemicals, *Chemical Engineering*, p. 156, April 18th, 1960
11. T. S. RICKETTS, High pressure coal gasification plants in Scotland and abroad, *J. Inst. Fuel.*, **34,** 177, 1961

CHAPTER 3

Properties of Liquid Fuels

MANY physical and chemical tests are used to determine the properties of liquid fuels, and the significance of these properties to the practical applications of the different types of fuels will be described in this chapter. The particular methods of test are described elsewhere,[1, 2, 3] and British Standard Specifications are available for most liquid fuels.

Density and Specific Gravity

The density and specific gravity of liquid fuels are used to convert volumes to weights and also, in conjunction with other properties, to evaluate petroleum products. Specific gravity, rather than density, is more commonly determined, and is defined as the weight of a given volume of oil to the weight of the same volume of water at a fixed temperature. Oils expand on heating and it is important to relate specific gravity to temperature. It is usual to measure this property at 60°F in relation to water at the same temperature, hence "specific gravity, 60°F/60°F".

The API scale is sometimes used in place of specific gravity, being related as follows:

$$°API = \frac{141 \cdot 5}{\text{specific gravity } 60°F/60°F} - 131 \cdot 5$$

The change in specific gravity with temperature depends on the coefficient of expansion of the particular oil, an approximate gravity correction for heavy petroleum fuels being 0.00035°/F.

While weight is an additive property and volume is not, for most practical purposes it can be assumed that no expansion or contraction takes place on blending, and the errors incurred when calculating the specific gravity of blends on this basis are seldom large.

For petroleum oils the specific gravity is generally less than unity, but for most coal tar fuels the gravity is greater than unity. The calorific value of petroleum oils is closely related to specific gravity according to the U.S. Bureau of Mines equation

$$CV_{\text{gross}} = 22,320 - 3780d^2 \text{ B.T.U./lb, where } d = \text{specific gravity, } 60°\text{F}/60°\text{F}$$

Other properties which can be qualitatively indicated by specific gravity include the aromatic content or C/H ratio of light distillates, and the burning quality of kerosines.

Viscosity

Viscosity is an important property of all liquid fuels, since it is a measure of their resistance to flow and thus has a large effect on the rate of flow through pipes and other items of equipment, on the flow of kerosine up a wick, on the atomization of fuel oils, and on the performance and wear of diesel pumps.

Viscosity can be defined as the resistance to movement between two layers of liquid moving with a relative velocity between one and the other. Absolute viscosity is measured in poises, a poise being the force required to move one square centimetre of plane surface at the relative rate of one centimetre per second to another plane surface, parallel to the first and separated one centimetre apart by a layer of the liquid. The kinematic viscosity is the ratio of the absolute viscosity to the density, the metric units being stokes or centistokes.

The viscosity measurements most often carried out are of the kinematic type, utilizing the fact that at very low speeds the resistance of a liquid to flow is proportional to the viscosity. Thus if the times taken for two liquids to flow through the same apparatus under the same conditions are known, together with the viscosity of one of the liquids, then the viscosity of the second can be obtained from the ratio of the two times. This is the basis of most standard methods of viscosity determination. In Britain, two methods are in common use: the "kinematic viscosity" method is based on the rate of flow in a glass U-tube viscometer; the "Redwood viscosity" is based on the rate of efflux from a metallic cup. Of these the kinematic determination is more precise and accurate, and is tending to supersede the Redwood viscosity for fuel oils.

Until recently it has been the practice to quote fuel oil viscosity in Redwood No. 1 seconds at 100°F, or for the more viscous oils in Redwood No. 2 seconds, the Redwood No. 2 cup having a larger orifice which allows the sample of oil to flow out in approximately one-tenth the time taken in the Redwood No. 1 apparatus. Owing to the difficulty experienced with waxy fuel oils in the determination at 100°F, and also in order to express viscosities in a universally accepted unit, it has now become the practice to quote kinematic viscosities at 122°F for the heavier fuel oils. Tables are available for the conversion of Redwood seconds to centistokes, or to the Continental Engler degrees, or the American Saybolt Universal seconds, or Saybolt Furol seconds, the latter being used in the U.S. for the more viscous oils.

Temperature has a marked effect on the viscosity of most oils, an increase in temperature causing a decrease in viscosity. The extent of this temperature effect varies with different oils and the higher the viscosity the greater is the effect of temperature change upon it. The viscosity–temperature relationship is expressed by means of the Viscosity Index, which is derived by an arbitrary method which compares the change in viscosity obtained over the temperature range 100°F to 210°F with the change obtained over the same temperature range with a Pennsylvanian oil (high *VI*, good viscosity–temperature characteristic) and a Gulf Coast oil (low *VI*, poor viscosity–temperature characteristic). For fuel oils the effect of temperature on viscosity is important, since viscous oils should be heated to a temperature at which the viscosity is not greater than 5000 Redwood 1 seconds, under which condition the oil should be readily pumpable. Similarly the temperature required for the best atomizing viscosity of the oil needs to be known. Viscosity indices are not normally quoted for fuel oils, but an approximate viscosity–temperature relationship for standard grades of fuel oil is given in Fig. 12.

Viscosity is not an additive property and charts are available for estimating the viscosity of blends.

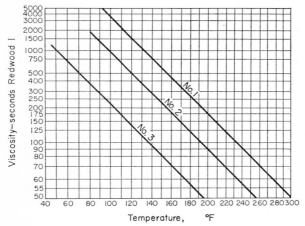

FIG. 12. Approximate viscosity–temperature relationship for petroleum fuel oils

Flash Point and Fire Point

The flash point is defined as the lowest temperature at which a product gives off just sufficient vapour to form an inflammable mixture with air under the conditions of the standard test. The fire point is the lowest temperature at which the oil will burn continuously when a flame is applied to its surface under prescribed conditions.

These empirically determined properties of oils are indirect measures of volatility, and serve as an indication of the fire hazards associated with their storage and use. Legal definitions of various types of fuel oil specify that the flash point should not be less than 150°F, and for other inflammable products which might have flash points of less than 150°F special regulations exist regarding safety in storage and handling. In the UK kerosine is required to have a flash point of 73°F minimum, but in practice the flash point generally exceeds 100°F.

The instruments used to determine flash points and fire points are of the closed and open cup types. In the former the oil is heated in a closed vessel until the temperature is reached at which the vapours in the air space above are sufficient to form an inflammable mixture and so ignite when a flame is applied. In the open apparatus the cup has no cover and the air above the liquid is in free contact with the surrounding atmosphere. The open flash point is thus a few degrees higher than the closed flash point, and the apparatus sometimes suffers from the disadvantage that traces of very volatile material, which could be detected by the closed flash apparatus, might be undetected in the open flash method. If the heating is continued beyond the flash point, a temperature will be reached at which the oil burns continuously on application of a flame. The temperature at which combustion is continuous is known as the fire point.

Flash points of such volatile products as gasolines are not usually determined, since these would be below normal atmospheric temperature and are termed "open flash points". For kerosines the Abel closed apparatus is used, whereas for gas oils and fuel oils the Pensky–Martens closed apparatus is most commonly used.

Pour Point and Cloud Point

In addition to viscosity, the flow of oil is affected by its physical state, that is, whether it is a liquid, a semi-solid, or a solid. Thus an increase in resistance to flow can result from an increase in viscosity or from the crystallization of wax from the oil.

The pour point of an oil is defined as "that temperature which is 5°F above the temperature at which the oil just fails to flow when cooled under prescribed conditions". The pour point gives an indication of the minimum temperature at which the oil will flow and is an important property of fuel oils and lubricating oils. In the case of heavier fuel oils, heating facilities are usually necessary in order to maintain the oil above its pour point and thus ensure satisfactory flow of the fuel. Failure to flow is generally due to the separation of wax from the oil, but in the case of very viscous oils may be the effect of viscosity alone. The result of the pour point test is affected by the shape of the vessel, by the pretreatment of the oil, by the quantity of oil used, and by the rate of cooling.

The cloud point of an oil is defined as "the temperature at which a cloud or haze appears when the oil is cooled under prescribed conditions". The cloud may be due to separated waxes or to water coming out of solution in the oil. The main use of the cloud point is to give an indication of the lowest temperature at which an oil can be used without causing blockage of equipment, in particular filters.

The pour point can often be markedly reduced by the use of pour point depressants, which form a coating around the wax crystals and so inhibit their growth and alter their shape to prevent the creation of a structure in the oil. These depressants modify the nature of the wax crystals but do not have much effect on the quantity of wax which has separated from the oil, so that the addition of depressants affects the cloud point much less than the pour point. Asphaltic substances act as pour point depressants, so that most residual fuel oils contain natural depressants.

Carbon Residue

Carbon residue gives a measure of the quantity of solid deposits obtained when medium and heavy fuel oils are subjected to evaporation and pyrolysis at elevated temperatures. The bulk of the oil will evaporate, but the heavier and more complex material will decompose and form carbonaceous deposits. This property is of importance for oils used in gas making, in compression ignition engines, in burners, and for lubricating oils used in internal combustion engines. The carbon residue tests in use are the Conradson and the Ramsbottom, the latter being used mainly for lubricating oils whereas the former is generally used for fuel oils. The Ramsbottom test produces the more accurate and reproducible results, and over the range 1 to 10% carbon residue the two tests are approximately related by Ramsbottom = 0·8 Conradson. For distillate fuels, such as gas oil, which have a low carbon residue, the test is carried out on the 10% residue after distilling off 90% of the sample.

Carbon residue is an important factor in the choice of fuels for compression ignition engines. For high speed diesel engines having short combustion periods, gas oils having a Conradson carbon residue of not more than 0·1% on a 10% residue are recommended. The nozzles for such engines are not usually water cooled, and the formation of carbonaceous deposits on the tips leads to poor operation and loss of power. The larger medium or slow speed diesel engines have a longer combustion time and these engines, particularly those having adequately cooled spray nozzles, can operate on fuels having much higher carbon residue, in some cases up to 12%. The nature of the deposition is of particular importance for diesel fuels, since fluffy deposits are easily removed by the exhaust stroke whereas hard deposits have a greater effect on the performance of the engine.

The significance of the carbon residue for oils used in gas making depends largely on the type of process used. For example, in the Jones process it is essential to produce solid carbon and the fuel used must have a Conradson carbon of 6–11%. On the other hand, the Segas catalytic process is capable of gasifying fuels having negligible, or as much as 12% carbon residue. A fuel of low carbon residue is desirable for the carburetted water gas process, but by the application of certain modifications such plants can be made to operate satisfactorily on residual fuels of higher carbon residue.

Pressure jet and blast atomizing burners are virtually insensitive to the carbon residue of the fuel, but vaporizing burners are greatly affected by deposits of carbon. Kerosines are the fuels mostly used for vaporizing burners and their carbon formation is often indicated by means of the wick-char burning test, the carbon residue being too low to have any significance. Gas oils can be used in the pot type vaporizing burners, and for this purpose the Conradson carbon should not exceed 0.05% on the 10% residue.

Burning Tests

The burning quality of kerosine is empirically assessed by means of the Smoke Point Test and the IP Burning Test. The results obtained from these tests enable the kerosine to be evaluated as an illuminant or as a fuel for space heaters, boilers, cookers, incubators, and so on, even though the type of appliance used in the test differs widely from those in the appliance being used.

The deposit-forming tendency of kerosines is too low to be measured by means of the carbon residue test, and it is therefore predicted from the char formed on the wick during the 24 hour IP burning test. Incrustations always form on the wick, some being soft and cause the flame size to increase, some being hard and enclose the surface of the wick to cause the flame size to decrease, whereas the worst type are irregular "mushrooms" which cause the flame to become distorted and usually smokey.

The smoke point test indicates the height of flame which can be obtained without the formation of smoke, and hence provides a measure of the illumination likely to be obtained from a particular kerosine in a particular wick-fed lamp.

Both the char value and the smoke point are related to the type of appliance used and also to the chemical and physical properties of the kerosine. In general, aromatic hydrocarbons increase the char formation and decrease the smoke point of a kerosine, and sulphur compounds are thought to increase the char problem.

Yellow flame wick-fed appliances, such as railway signal lamps, incubator lamps, and some types of convector space heaters, require a kerosine of low char value since they are required to burn for prolonged periods without attention. Blue flame appliances, as used in boilers and space heaters, are less sensitive to kerosine quality, but nevertheless perform best on a kerosine of low char value.

Sulphur Content

With the exception of carbon and hydrogen, sulphur is undoubtedly the most important element present in liquid fuels, since it has some effect on all possible applications. These effects are invariably bad, causing corrosion in some products, bad odours in others, and possibly having an influence on the char value of kerosines, the lead response and colour stability of gasolines, and the stability of gas oils.

When comparing the sulphur content of various fuels the basis should be sulphur per unit of heat output, not as is frequently encountered, sulphur per unit weight of fuel, regardless of the calorific value.

Liquefied petroleum gases contain sulphur as H_2S and methyl mercaptan. H_2S has to be removed completely and the total sulphur content of the gases when sold must not exceed 0·02% by weight. Liquid fuels used for gasmaking should have as low a sulphur content as possible in order to reduce the cost of H_2S removal from the product gases, although in fact the purification costs are relatively small compared with the cost of the feedstock.

The removal or conversion of the various forms of sulphur from gasolines, kerosines, and diesel oils has been described in Chapter 1, and is necessary because of the corrosive nature of the element as it occurs in engine fuels and also because the emission of sulphur oxides is undesirable from equipment which burn these fuels.

In the case of heavy distillate and residual fuels there is not as yet a commercial process for the removal of sulphur. There are three possible effects of sulphur when such fuels are burned, namely the contamination of stock by sulphur in direct fired furnaces, the corrosion of low temperature surfaces due to condensation of sulphuric acid from the flue gases (Chapter 6), and the pollution of the atmosphere with sulphur oxides. For certain processes in which the contamination of stock must be minimized, the only solution is to select low-sulphur fuels. For example, the production of low sulphur steels (0·02%S) in open hearth furnaces fired with fuel oil of 1·5% sulphur is difficult, and in this respect coal tar fuels with a maximum of 1·0% sulphur are preferred to the majority of fuel oils derived from petroleum. The mechanism of sulphur pick-up by the steel is not fully understood, and the evidence relating sulphur content of fuel to sulphur content of steel is limited.

The determination of sulphur ([4]) is carried out by the use of X-rays and by means of the lamp, quartz tube and bomb combustion methods. The tendency to cause corrosion is assessed by the copper strip test.

Ash

For most petroleum fuel oils the ash content varies from nil in gas oil up to 0·1% in residual fuels, whereas for coal tar fuels the ash content increases from about 0·05% for CTF 50 to 0·3% in CTF 250. In general, the ash content increases with increasing viscosity. Unlike petroleum fuel oils, coal tar products contain no vanadium. The presence of this element, together with sodium in petroleum oils plays an important part in certain combustion systems. In glass tanks vanadium can lead to discoloration of the glass, and since this element cannot be removed from the oil, low vanadium fuels have to be obtained from selected crudes. In gas turbines the use of residual fuels results in corrosion and fouling at temperatures of about 1200°F and above, this temperature being about the melting point of the complex compounds of vanadium, sodium, oxygen and sulphur formed from the ash. The efficiency of a gas turbine improves as the operating temperature is increased, so that the limitation imposed by the inorganic constituents of the fuel oil is serious. There are possibilities that the fouling and corrosion can be reduced, either by the control of combustion conditions to give some free carbon which modifies the softening characteristics of the vanadium compounds, or by the use of inhibitors. Corrosion due to vanadium has not occurred in boilers, since metal temperatures approaching 1200°F have not been encountered as yet. However, sodium if present in sufficient amounts, as sometimes occurs in contaminated marine fuels, can cause trouble due to bonding of ash constituents on superheaters and the attack on refractories. [5]

The presence of ash in diesel fuels contributes to fuel pump, piston, and ring wear, to engine deposits, and to poor operation of the smaller injectors.

Stability and Compatibility

Fuel oils and diesel fuels which contain cracked products tend to form gums and sediments when in contact with air and water. This instability can cause trouble due to deposition in storage tanks, lines, and filters, and where blending components of poor compatibility are used the deposition is more pronounced. Untreated cracked distillates deteriorate faster than straight-run distillates, and when these two types of products are mixed there is a strong tendency for the straight distillates to precipitate gums from the cracked products. Additives can be used to improve stability and also to inhibit rust formation.

Water

The solubility of water in hydrocarbons is low (0·005 to 0·05% by weight) and varies with the type of hydrocarbon and with the temperature. Water up to 1% may be encountered as a dispersion in residual fuel oils, but as long as it exists in a fine state of division, gives rise to no practical difficulties, except that the salt in sea water can cause damage to refractories. When oils are preheated the separation of water is reasonably rapid.

Calorific Value

For petroleum fuel oils the gross calorific value ranges from about 18,000 B.T.U./lb for the heaviest to almost 20,000 B.T.U./lb for light distillates. The net calorific value differs from the gross by approximately 1050 B.T.U./lb for the whole range of petroleum fuel oils.

For coal tar fuels the calorific value on a weight basis is appreciably less than for petroleum fuels, and the gross value varies from about 16,100 B.T.U./lb for CTF 250 to 16,700 B.T.U./lb for CTF 50. It should be noted that, whereas calorific value on a weight basis is more for petroleum fuels than for coal tar fuels, on a volume basis the calorific value is greater for the coal tar fuels, this being due to the large difference in specific gravity between the two types of fuels.

Engine Tests and Volatility Tests

These tests and properties of fuels are dealt with in Chapters 4 and 5.

References

1. *Standard Methods for Testing Petroleum and its Products*, Institute of Petroleum, 1960
2. *ASTM Standards on Petroleum Products and Lubricants*, ASTM, 1960
3. *Standard Methods for Testing Tar and its Products*, Standardization of Tar Products Tests Committee, 1957
4. M. C. FRANKS and R. L. GILPIN, *J. Inst. Pet.* 48, 226, 1962
5. W. A. WALLS and W. S. PROCTOR, *J. Inst. Pet.* 48, 105, 1962

CHAPTER 4

Fuels for Spark Ignition Internal Combustion Engines

THE main fuels used in spark ignition engines are motor gasoline, aviation gasoline and vaporizing oil. In this type of engine the fuel is mixed with air to form an inflammable mixture, compressed in one of the engine cylinders and ignited. The mixture burns and the resulting increase in pressure moves the piston and provides the power stroke.

The efficiency of such an engine depends on its design and on the properties of the fuel used. The main factor in the case of engine design is the compression ratio, i.e. the ratio between the volume of the cylinder before and after compression of the mixture, since the higher the compression ratio the greater the efficiency. Compression ratios are limited, however, by the characteristics of the fuels used since for each fuel there is a maximum useful compression ratio above which "knocking", or "pinking", occurs, i.e. detonation in the engine cylinders. This phenomenon causes a decrease in power output, overheating and, if prolonged, damage to the engine.

One of the main requirements of a fuel for this type of engine therefore, are satisfactory combustion characteristics to prevent "knocking".

Combustion Characteristics

In the cylinders of a spark ignition engine the combustion of an air/fuel mixture results in the liberation of energy and a rapid build-up of pressure. During normal combustion the burning occurs in a narrow flame front which travels through the mixture

FUELS FOR SPARK IGNITION ENGINES

at a moderately fast rate. The pressure rise is thus a gradual one and there are no surges or vibrations in the cylinders. The rate of travel of the flame through the mixture under these conditions is of the order of 50 ft/sec, but varies with the design of the engine cylinder, engine speed, and the composition and temperature of the air/fuel mixture. During the combustion a slow oxidation of the unburnt gases takes place in front of the advancing flame. The oxygen present combines with some of the hydrocarbon molecules giving them a high energy content and these molecules then react with adjoining molecules so that a chain reaction is initiated. The amount of unstable compounds thus formed is very small, and the combustion proceeds in a smooth manner.[1]

Engine "Knock"

Under certain engine conditions combustion will commence in the normal manner then suddenly the remaining unburnt fuel will ignite spontaneously. In this case the chain reactions described above proceed very rapidly so that a high concentration of unstable products is formed which propagates a flame at very high velocities, up to 1000 ft/sec. Pressure waves are then set up which vibrate against the cylinder walls giving rise to a "knocking" sound.[1]

Surface Ignition

Abnormal combustion can also occur in high compression engines as a result of "surface ignition effects". In this case high engine temperatures result in hot engine deposits which may ignite the cylinder charge at the wrong time. This causes sudden pressure surges through the engine cylinders and can result in mechanical damage to the engine bearings.

Factors Affecting Engine "Knock"

As already stated, knocking in an internal combustion engine is governed by the chemical composition of the fuel charge and the physical characteristics of the engine. In the case of the fuel the most important factor is the properties of the individual hydrocarbons. Hydrocarbons which tend to react with oxygen under the conditions which prevail immediately prior to combustion will detonate readily. Research work on the behaviour of different types of hydrocarbons in this type of engine has shown that in general straight chain paraffins have a marked tendency to detonate whereas branched chain paraffins and aromatics have good anti-knock properties. Olefines and naphthenes have intermediate properties.[2]

For a given engine, low working temperature, low compression ratio, high engine speed, absence of cylinder deposits and dead spaces in the cylinder head, are all factors which reduce the tendency to knock. If the compression ratio is varied, all other conditions remaining constant, knocking will occur at different ratios for different fuels. Thus in any engine it is possible to find the highest useful compression ratio for a given fuel. Since the value for any fuel will be different in different engines, tests are made in standard engines operating under standard conditions to compare the anti-knock values of different fuels. An engine developed by the Coordinating Fuel Research Council in America is used by all the Oil Companies for this purpose and the test is designated as the C.F.R. Engine Test. This engine has a single cylinder with an adjustable head for varying the compression ratio, and operating conditions such as engine speed, coolant temperature, spark advance, fuel mixture temperature and valve clearances are rigidly controlled. Knock intensity is measured with a detonation meter which detects the rate of change of pressure in the engine cylinder.

Octane Number

The anti-knock quality of a fuel is usually expressed in terms of Octane Number. In the test engine the knocking properties are compared with blends of two hydrocarbons, n-heptane which detonates readily at low compression ratios, and iso-octane which is highly resistant to detonation. These hydrocarbons are designated 0 and 100 respectively in the Octane Number scale and are blended to give a material which will match the knock characteristics of the fuel being tested. The percentage of iso-octane in this blend is termed the Octane Number of the fuel under test. Reference fuels with octane numbers in excess of 100 are prepared by adding tetraethyl lead, TEL (an anti-knock additive), to pure iso-octane. The Octane Number of a fuel under test is then reported as 100 + the quantity of TEL added to iso-octane to match the anti-knock quality of this fuel. Thus a typical result would be 100 + 1·5 (ml TEL per gallon), or iso + 1·5. Alternatively an Octane Number above 100 may be obtained from Tables which relate "knock value" (ml TEL per gallon), to Octane Number.

Motor Gasolines

The first motor fuels were straight run gasolines, i.e. light fractions obtained from the distillation of crude oil. Engine compression ratios at this time were about 4 to 1, but as more efficient and powerful engines were developed, new gasolines of improved quality were manufactured. Engines in pre-war cars had compression ratios of about 6 to 1 and operated on gasolines with Octane Numbers (Research) of the order of 70. Today compression ratios have increased to up to 10 to 1 and the Octane Numbers of gasolines up to 101. To obtain these higher octane gasolines, new manufacturing processes have been developed in the petroleum refineries and today gasolines contain some, if not all, of the components shown in Table 16.

Motor gasolines are mixtures of hydrocarbons with an approximate boiling range of 30–200°C. Small quantities of sulphur compounds and various additives are also present and they may also contain components such as benzole and alcohol. The performance of these gasolines in a spark ignition engine is governed by the properties of the hydrocarbons present in the gasolines. Thus the chemical nature of the hydrocarbons should be such that at the conditions prevailing in the engine cylinders

Table 16

Present Day Motor Fuel Components

Component	Typical Octane Nos.	
	Motor method	Research method
Natural gasoline	60	60
Straight run gasoline	55	55
Alkylate	90	92
Catalytic reformate	88	95
Catalytic polymer gasoline	83	97
Isomerized paraffins	80	80
Cat. Cracked Gasoline	80	93

they will burn smoothly without detonation and leave no deposits. In addition, no corrosive or gum forming constituents should be present and the volatility characteristics should be such that the engine will start and perform satisfactorily under all engine operating conditions.

The properties of gasolines marketed currently in the United Kingdom are shown in Table 17.[3]

Volatility

For a spark ignition engine to start and perform satisfactorily a suitable fuel/air mixture must be introduced into the engine cylinders under all the different engine operating conditions. Mixtures of the required composition are provided by a carburettor in which small quantities of fuel are drawn from a jet by a current of air passing through a venturi in which the jet is situated. The fuel is then atomized and partly vaporized, and by the use of equipment such as a choke and compensating jets the fuel composition can be varied to meet the engine requirements. For example, to enable a cold engine to start the fuel/air mixture is enriched to be easily ignited under cold start conditions. This is

TABLE 17

Range of Properties of Motor Gasolines Marketed in the U.K.
(first two months of 1961)

Grade of gasoline	Regular		Premium		Super	
	Min	Max	Min	Max	Min	Max
Octane No.—Research	83	84	95	99	100	101
Motor method	79	80	84	88	88	90
Lead content/US gal						
—T.E.L. ml.	1·2	1·7	1·0	2·3	1·2	2·0
—Lead g	1·2	1·79	1·05	2·43	1·26	2·11
Reid vapour pressure, psi	6·8	11·8	8·0	11·0	8·1	10·3
Specific gravity, 60/60°F	0·704	0·717	0·720	0·753	0·739	0·759
Total sulphur % wt	0·05	0·08	0·05	0·15	0·04	0·09
Distillation °C						
I.B.P.	25·5	41	27	45	25	31·5
10% vol at	45	57	46·5	67	42	52
50% vol. at	80·5	91	86	113	95	110·5
90% vol at	138·5	175	153	170	155	172
F.B.P.	170	206	180	205	179	204

achieved by using the choke. When the engine becomes hotter more fuel is vaporized as the mixture approaches the heated engine and the use of the choke is no longer necessary.

The volatility characteristics of the fuel are extremely important and laboratory tests are carried out to evaluate these. The main tests are distillation and vapour pressure, the results of which can be correlated with engine performance. The vapour forming characteristics of a gasoline under conditions existing in engine fuel systems can be evaluated by the vapour/liquid ratio test. This shows the volume of vapour formed per unit volume of liquid at a series of different temperatures and indicates the tendency of the fuel to "vapour locking". (Vapour lock in an engine occurs when the formation of vapour in the fuel feed system prevents the fuel pump from supplying liquid fuel to the carburettor.)

The various volatility problems associated with engine performance and the portion of the boiling range of a fuel related to these are shown in Table 18.[4] Some of these problems are seasonal but each one can be controlled by varying the front, middle or back end of the distillation curve.

TABLE 18

Volatility Problems Related to a Certain Portion of the Boiling Range of a Fuel

Engine problem	Front end (10–30% point)	Mid range (50–70% point)	Back end (90% point)
Cold starting†	(1)		
Evaporation loss	(1)		
Vapour lock*	(1)	(2)	
Hot starting*	(1)	(2)	
Hot idle*	(1)	(2)	
Carburettor icing†	(1)	(1)	
Warm-up†	(2)	(1)	(2)
Acceleration		(1)	(2)
Calorific value			(1)
Oil dilution			(1)
Cleanliness			(1)

(1) Prime factor
(2) Lesser but significant factor
* Hot weather problems
† Cold weather problems

Front end characteristics.—These govern the ease of starting of an engine under different temperature conditions, the tendency of a fuel to vapour lock, and the evaporation loss. By increasing the proportion of light hydrocarbons in a fuel the ease of starting is improved but evaporation losses and vapour locking problems increase. Therefore, the front end characteristics, or 10% distillation point are chosen to permit, as far as possible, ease of starting and yet prevent vapour locking. Since these factors are related to atmospheric conditions, the amount of light materials in a gasoline is varied with the season. Thus under winter conditions the fuel must contain sufficient light hydrocarbons to permit satisfactory starting at low temperatures, but with higher ambient temperatures during the summer, the percentage of these hydrocarbons may be reduced. The 10% distillation point and Reid vapour pressure maximum specification points for typical winter and summer gasolines are 55°C and 12 psi, and 65°C and 10 psi respectively.

Mid boiling range.—The two main factors affected by this range of the distillation curve are the warm-up performance of an engine and acceleration. Warm-up performance, or the initial operation of the engine after starting, may be defined as the time or distance travelled before the engine operates smoothly without use of the choke. It is directly affected by the ambient temperature, and is controlled mainly by the mid-range volatility of the fuel used and to a smaller extent by the 90% distillation point.

For good acceleration and warm-up performances the 50% distillation point of a fuel should be kept below about 105°C.

Carburettor icing occurs under conditions of low temperature and high relative humidity, and with highly volatile fuels. The evaporation of such a fuel in the carburettor lowers the air temperature and results in the formation of ice. This tends to restrict air and fuel flows and causes the engine to stall. To avoid ice formation the 50% distillation temperature of a fuel should be kept as high as possible but since this would give poor warm-up and acceleration characteristics this problem is avoided by using anti-icing additives. Carburettor icing is mainly experienced during engine start-up but can also occur when an engine is operating under part throttle conditions. In both cases it is often related to the design of the carburettors.

Back end characteristics.—These mainly affect the heat content of the fuel, crankcase oil dilution and engine cleanliness.

In general, the heavier hydrocarbons have higher heat contents than the lighter ones and, on a volumetric basis, a heavy fuel will contain more energy per gallon than a light fuel. However, the actual differences in car performance are very small.

Crankcase oil dilution results mainly from high boiling components in a gasoline. Unburnt fuel remaining in the engine cylinders washes the lubricant from the cylinder walls and works its way past the piston rings into the crankcase. In addition to the volatility characteristics of a fuel, atmospheric conditions and the mechanical condition of the engine will also affect oil dilution. Excessive dilution in this way reduces the lubricating properties of the oil and results in increased engine wear. Engine cleanliness also depends on the amount of heavy hydrocarbons present in the fuel since these materials will form deposits in the engine cylinders.

Engine Testing

The two methods of test used for rating motor fuels are the Motor Method and the Research Method.[5] The Motor Method is the more severe test (engine speed 900 rev/min, mixture temperature 300°F) and corresponds fairly closely to the operation of car engines at high speeds and under heavy loads. The Research Method (engine speed 600 rev/min, air inlet 120°F) gives good agreement with engines operating under mild conditions such as low engine speed.

Because of the large number of automobile engines in use today, and the widely different operating conditions of these engines, neither of the above tests will accurately predict the anti-knock quality of a fuel in a car, i.e. its road performance. Correlations between laboratory and road ratings have been made however, and the difference between the Research Octane Number and the Motor Octane Number, termed the sensitivity of a fuel, gives a good indication of the behaviour of a fuel in service. Fuels tend to road rate intermediate between their Motor and Research numbers and modern cars tend to rate nearer the Research Octane Number. In general, low sensitivity fuels will road rate better than high sensitivity fuels of the same Research Octane

FUELS FOR SPARK IGNITION ENGINES

Number, although the distribution of octane number through the boiling range of a gasoline can also have a considerable effect on road rating.

Gasoline Additives

Small amounts of certain chemicals reduce the tendency of gasolines to knock, i.e. they improve the octane rating of the gasolines. The action of these compounds is to react with those activated oxygenated intermediate compounds whose decomposition products lead to knocking combustion. Of the large number of additives experimented with the most effective in this field has been TEL* and about 2 ml/gallon are currently added to gasolines in the United Kingdom. Larger amounts, up to 4 ml/gallon, are used in the American continent and in parts of Europe. The increase in octane number resulting from the addition of a given quantity of this material, termed the lead susceptibility, is not uniform but depends on the composition of the gasoline. Sulphur compounds should be reduced to a minimum since these reduce the lead susceptibility.

TEL is invariably used with a scavenger to aid the removal of the reacted lead compounds from the engine cylinders and reduce the formation of deposits. Volatile halides such as ethylene dichloride and dibromide are used and these form volatile lead halides.

A recent development in this field has been the use of tetramethyl lead, TML, which has been shown to be superior to TEL in some super grade gasolines.

Other types of additives which may be used in gasolines are given in Table 19.[6]

Aviation Gasolines

Aviation gasolines are, in general, similar to motor gasolines except that they have a slightly narrower boiling range, about 40 to 170°C, and a lower vapour pressure, 7 psi maximum. Until

* Tetraethyl lead.

Table 19
Gasoline Additives

Additive	Prevents	Typical chemicals
1. Combustion chamber deposit modifiers	Spark plug fouling	Organic phosphates.
2. Anti-oxidants	Gum formation	Phenols, amines.
3. Anti-rust agents	Rust due to condensation or contamination	Unsat. organic acids, orthophosphates.
4. Copper deactivators	Gum formation catalyzed by metals	Amines, aminophenols.
5. Fuel induction system cleanliness agents	Deposition of contaminants in this system	Heavy petroleum oils. Organic silicone compounds.
6. Carburettor anti-icing agents	Ice formation from water in the air	Glycols, alcohols, amine phosphates.
7. Dyes	—	Azo- and orthoquinone type.

recently the anti-knock quality of these fuels was higher than that of motor fuels but today the Octane Numbers of the best quality motor spirits are in the aviation gasoline range.

Gasolines of this type are made up of specially selected straight run and synthetic components such as iso-pentane, obtained from straight run gasoline by superfractionation, alkylates and aromatics. The latter are obtained from reformates by distillation or solvent extraction. The main requirements in some typical aviation gasoline specifications are shown in Table 20.[7]

As in the case of motor gasolines, octane rating and volatility are the two most important properties of these gasolines. The anti-knock properties are again measured in a standard CFR single cylinder engine but operating under different conditions from those used for testing motor gasolines. In this case two ratings are obtained. One simulates operation under take-off conditions, when maximum power output is required and an enriched fuel mixture and supercharged engine are employed,

TABLE 20
Typical Aviation Gasoline Specifications

Grade		80	91/96	100/130	115/145
Colour		Colourless	Blue	Green	Purple
Tetraethyl lead	ml/gal max	Nil	5.5	5.5	5.5
Distillation:					
Evaporated at 75°C	% vol	10–40	10–40	10–40	10–40
Evaporated at 105°C	%vol min	50	50	50	50
Evaporated at 135°C	% vol min	90	90	90	90
FBP °C max		170	170	170	170
Sum 10% + 50% evap. temps. min		135	135	135	135
Reid vapour pressure	psi	5.5–7.0	5.5–7.0	5.5–7.0	5.5–7.0
Sulphur	% wt max	0.05	0.05	0.05	0.05
Freezing point	°C max	−60	−60	−60	−60
Gum existent	mg/100 ml max	3	3	3	3
Gum accelerated	mg/100 ml max	6	6	6	6
Octane No.—clear, motor method F2 min		80	90	99	—
—Aviation method, weak mixture rating, min		—	91	100	115*
—Aviation method, rich mixture rating, min		—	96	130*	145*

* Performance Number

and the other under cruising conditions. In the latter case fuel economy is the main consideration and weak fuel mixtures are used. The tests are designated Aviation Method—rich mixture rating and weak mixture rating respectively.

Weak Mixture Rating

This test is similar to those carried out for motor gasolines except that the engine speed is higher, 1200 rev/min, and the anti-knock quality is determined by measuring the high temperatures which occur in the engine cylinder by means of a thermocouple.

Rich Mixture Rating

This test is entirely different in that a rating is obtained by comparing the "knock limited power" developed by the fuel and comparing it with blends of reference fuels. The "knock limited power" is defined as the maximum power the engine will produce under a given set of conditions without "knocking", although in an engine test the rating is determined at the least knock that an operator can recognize. The test is carried out with an engine speed of 1800 rev/min and a fixed compression ratio (7 to 1), the fuel/air ratio and manifold pressure (amount of supercharge) being varied until knocking occurs.

In both tests the same reference fuels are used as in the testing of motor gasolines.

Performance Number

Anti-knock values of aviation gasolines are quoted as Octane Numbers for ratings up to 100 but for higher values the ratings in terms of leaded iso-octane are converted to Performance Numbers. The Performance Number of a fuel is defined as the percentage gain in "knock limited power" developed by a typical supercharged aviation gasoline engine when operating on that fuel as compared with that obtained when operating on iso-octane. Thus in such an engine a gasoline with a Performance Number of 130 would develop about 30% more knock free power than it could with iso-octane. The Performance Number scale was developed following a series of full scale tests in which the percentage gain in power in typical aviation engines was measured when operating on iso-octane containing various amounts of TEL as compared with unleaded iso-octane.

Anti-knock ratings, as a guide to engine performance, are more critical in the case of aviation gasolines than for motor gasolines since the excessive noise of an aircraft engine could allow knocking and possible mechanical damage to pass undetected. In addition, because of the much higher operating temperatures, maloperation could result in extremely high localized temperatures and possibly melting of metal parts.

The various specification points in the boiling range of aviation gasolines are related to the same volatility problems as for motor gasolines. Thus the percentage distilling to 75°C controls ease of engine start and vapour locking, and the 50% evaporated point the warm up performance and acceleration. The 90% point controls the back-end volatility, to ensure uniform fuel distribution to all cylinders and freedom from lubricating oil dilution.

The vapour pressure is maintained below 7 psi as an additional precaution against vapour locking. Other important properties of aviation gasolines are a sufficiently low freezing point to prevent icing difficulties at high altitudes and the absence of gum, since accumulation of this latter material could cause engine failure.

Some of the additives listed previously are also used in aviation gasolines although in the case of TEL, because of the different engine operating conditions, ethylene dibromide only is used as a scavenger.

Vaporizing Oil or Power Kerosine

Vaporizing oil is a fuel manufactured for spark ignition type engines but is outside the normal gasoline range and, therefore not subject to the taxation levied on gasolines. It is used in tractor engines of low compression ratio, about 4·5 to 1.

The boiling range is similar to that of kerosine and it is manufactured by blending straight run distillates with sufficient aromatics, obtained by the solvent extraction of kerosine, to give a material with an Octane Number of about 50 (Motor Method). Because of the absence of readily volatile materials, tractor engines are fitted with two carburettors, and the engine is started on ordinary motor gasoline and then switched to vaporizing oil for normal operation.

The development of diesel tractors has resulted in a marked decrease in the demand for this type of fuel.

Alcohols

The lower boiling point alcohols, i.e. methyl and ethyl, can be used in internal combustion engines, their Octane Numbers (Motor Method) being 98 and 99 respectively. In spite of these high ratings, however, they do not compare favourably with petroleum gasolines because they have much lower heat contents (resulting in poor fuel consumption figures) and are generally more costly.

They are normally used in blends with petrol (about 10 to 20%) and mainly in countries where they are obtained as by-products in other manufacturing process. In these countries the use of alcohols in motor spirit blends is often compulsory. Their main value is in upgrading low octane number materials such as straight run gasolines.

References

1. H. A. BEATTY and G. EDGAR, *Science of Petroleum*, Vol. 4, p. 2927, Oxford University Press (1938)
2. W. G. LOVELL, *Science of Petroleum*, Vol. 5, Part 3, p. 43, Oxford University Press (1955)
3. *World-wide survey of motor gasoline quality*, The Associated Ethyl Company Limited, London, April (1961)
4. G. H. UNZELMAN and E. J. FORSTER, *Petroleum Refiner*, **39,** 10, 109 (1960)
5. *I.P. Standards for Petroleum and its Products*, Part 2, Methods for rating fuels—Engine tests, October (1960)
6. Gasoline Additives, *Petroleum Refiner*, **38,** 9203 (1959)
7. W. A. PARTRIDGE, *Science of Petroleum*, Vol. 5, Part 3, p. 1, Oxford University Press (1955)

CHAPTER 5

Fuels for Compression Ignition Engines

THE most satisfactory fuels for compression ignition or diesel engines are petroleum distillates boiling in the gas oil range, and the majority of engines, in particular diesel engines in commercial vehicles, operate on a fuel of this type. For reasons of economy however, cheaper heavier fuels incorporating crude oil residues are used in some large diesel units in industry and in ships. The deciding factor in this case is what fuels the various engines can tolerate without excessive maintenance.

Compression ignition engines are mechanically similar to spark ignition engines but different methods are used to introduce the fuel into the engine cylinders and to ignite the air/fuel mixture. The fuel is injected directly into a cylinder containing a highly compressed air charge, and the heat of compression ignites the atomized fuel spontaneously.

Compression Ignition Engines

Compression ignition engines operate on either a two stroke or a four stroke cycle and a brief description of the latter operation is given to illustrate the fuel requirements of this type of engine. The four strokes in the thermal cycle are (1) Induction, (2) Compression, (3) Power and (4) Exhaust. An inlet valve and an exhaust valve are fitted in the top of each cylinder and the cycle commences with the piston at the top of the cylinder. During the induction stroke the inlet valve is open and air is drawn into the cylinder by the downward movement of the piston. At the completion of this stroke, the inlet valve closes and the piston travels upwards and compresses the air charge to about 600 psi, this compression raising the temperature of the air to about 600°C. Just before the completion of the compression stroke the injection of a spray of fuel commences and the high temperature of the air mass ignites the fuel. The burning of the fuel and the expansion of the gases moves the piston downwards and provides the power stroke. At the bottom of this stroke the exhaust valve opens and the products of combustion are removed from the cylinder by the second upward movement of the piston. Thus there is one power stroke for one complete cycle, and this corresponds to two revolutions of the engine crankshaft.

The high working pressure of compression ignition engines necessitates a more robust construction than for spark ignition engines, and the higher compression ratios of up to twenty to one give improved thermal efficiency, about 35% as compared with 25% for a gasoline engine. Engine performance, however, is related to the properties of the fuels used and to engine design, the two most important factors in the latter case being the injection equipment and the combustion chamber.

Fuel Injection

Injection equipment is designed so that the fuel injection begins at the right time (10 to 20° of crank angle before top dead centre) during the compression stroke and continues during the power stroke. In this way a constant pressure is maintained due to the combustion of the fuel. In addition this equipment controls the quantity of fuel which is injected into each cylinder and distributes the fuel spray through the cylinders in a suitably atomized form. Therefore the fuel should be sufficiently fluid at all operating temperatures to permit easy flow to the injection system. In addition its viscosity should be sufficiently low so that efficient atomization is obtained and no large droplets penetrate too far into the cylinders and impinge on the cylinder walls. The viscosity should not be too low however or fuel leakage past the fuel pump plunger might occur and result in a lowering of the amount of fuel injected.

Fuel injection was originally achieved by using compressed air but today air is not normally employed and the operation is termed "solid injection". A number of different types of injection equipment are used but these fall into two main categories, the Common Rail System and the Jerk System.

In the Common Rail System the injection nozzles to each cylinder are supplied by a common fuel line, and a fuel pump maintains a pressure of from 4000 to 10,000 psi in the line. The fuel injection valves are opened and closed mechanically. This system, although still in use in some large engines, has been virtually superseded by the Jerk System, where separate fuel pumps are used to deliver fuel to each cylinder during the injection period only.

There is a wide variety of different nozzles and combustion chamber designs. In the latter case the aim is to impart vigorous movement to the air charge and achieve an intimate mixture of fuel and air. In some cylinders this is obtained by using suitably disposed air inlet ports (air turbulence method) whereas others employ precombustion chambers to obtain a premixing of the fuel with a limited quantity of air (separate chamber type).[1] Some combustion systems have been developed which will burn either motor gasoline or diesel fuels.

Fuel Ignition

Following the injection of atomized fuel into the cylinders of a diesel engine, the oil absorbs heat from the air charge and attains the ignition temperature, which is dependent on the type of fuel. The period between the beginning of injection and subsequent ignition of the fuel is termed "ignition lag" or "ignition delay". Fuels which have a short ignition delay are defined as high ignition quality fuels, although delay times are also affected by the design and operating conditions of the engine. Engine factors which could affect ignition delay are compression ratio, fuel and air inlet temperatures, fuel injection time, air turbulence, and engine speed.

In extreme cases of ignition delay the fuel charge will not ignite. In less severe cases an excessive accumulation of fuel can occur prior to ignition so that when the mixture ignites there is a sudden pressure surge through the cylinders and detonation results. Changes in operating conditions which increase the temperature and pressure of the air charge at the end of the compression stroke will reduce the tendency towards "diesel knock" whereas similar changes in spark ignition engines increase engine "knock". The "ignition delay" of a diesel fuel should be as low as possible with the proviso that ignition does not occur before the completion of the compression stroke.

Fuel Combustion

After ignition, fuel is injected into the burning mixture and burns at the rate at which it is injected. Complete combustion is necessary to obtain maximum power output and minimum fuel consumption, and excess air is essential to burn the last droplets of injected fuel. Unless the quantity and turbulence of the air are sufficient, carbon deposition and the emission of partly oxidized hydrocarbons from the cylinders as a black, strong smelling smoke will result. Carbonaceous deposits around the fuel injection nozzles results in maldistribution of the fuel spray and increase fuel consumption, the effect being accelerated if nozzle temperatures are too high. Carbon particles also cause pitting of exhaust valves and lubricating oil sludging. Fuels with high carbon residues form deposits which tend to accumulate in the engine cylinders, and therefore the carbon residue of diesel fuels should be as low as possible, although low speed engines can tolerate larger amounts than the high speed engines because of the longer time available for combustion and because the burners have water cooled nozzles.

The viscosity characteristics of a fuel affect combustion in that this property influences the size of the fuel particles leaving the injection nozzles and consequently the time required for complete combustion.

FUELS FOR COMPRESSION IGNITION ENGINES

Characteristics of Diesel Fuels

A wide variety of compression ignition or diesel engines exists and these operate at widely different speeds and loadings. A general classification into three main groups can be made as shown in Table 21. The characteristics of the fuels used depends on such factors as engine size, speed and load ranges, and the frequency of the speed and load changes. Fuel quality is more critical the higher the engine speed and the smaller the engine, and in general high speed engines require fuels with lower viscosities and better ignition qualities than low speed engines.

TABLE 21

General Classification of Diesel Engines

Classiffication	Speed, rev/min	Operating conditions	Typical uses
High speed	900–4000	Frequent and wide variations in loads and speeds	Road transport. Diesel locomotives.
Medium speed	300–900	Relatively high loadings, fairly constant speeds	Large stationary engines. Auxiliary marine engines.
Low speed	Below 300	Sustained heavy loads, constant speeds	Marine main propulsion engines. Electrical power plant.

Fuels for high speed and the majority of medium speed engines are distillate gas oil fractions boiling in the range 180 to 360°C. Some medium speed engines and all low speed engines operate on heavier fuels, which are usually blends of gas oils with crude oil residues although residual fuels alone are used in some slow speed engines. The viscosity of these latter fuels may vary up to 3000 Redwood 1 seconds at 100°F, whereas the viscosity of high speed engine fuels is of the order of 40 Redwood 1 seconds at 100°F.

Centrifuges are employed in engines utilizing residual fuels to remove water and contaminants, and in addition these fuels are heated prior to injection into the engine cylinders to improve their pumping and viscosity characteristics.

Distillate fuels are comparatively free from the solid matter associated with residual fuels, but in all engines the fuels are filtered prior to pumping to the cylinders in order to reduce abrasion in the fuel pumps, injectors and cylinders. The acidity of diesel fuels should be negligible in order to avoid corrosion of the injection system.

Ignition Quality

The ignition quality of a fuel can be expressed in terms of a Cetane Number, which is based on the engine testing of the fuel, or in terms of various calculated indices, the most widely used being Diesel Index. In all cases the calculated results only give an approximation of the Cetane Number.

Cetane number.—This value is obtained by comparing the ignition characteristics of the fuel with blends of two reference fuels, these being cetane, which ignites readily, and alpha-methyl naphthalene, which is of poor ignition quality. The percentage of cetane in a blend having the same ignition delay as the fuel under test gives the Cetane Number of this fuel. The test is carried out in a standard single cylinder compression ignition engine. In the U.K. one of two methods is usually employed; an actual determination of the ignition delay (IP 41/60 Method A),[2] or the measurement of the air intake pressure to the engine cylinder at which the engine misfires (IP 41/60 Method B).[2] In both cases the results are compared with those observed on blends of the reference fuels.

In Method A the difference in time between the beginning of the fuel injection and the commencement of combustion, measured by the beginning of a rapid increase in cylinder pressure, is determined using a cathode ray oscillograph pressure indicator. In Method B an expansion chamber is fitted to the air inlet port to the engine and the air supply to this chamber is controlled by a throttle valve. This valve is gradually closed until a critical intake pressure is reached at which level insufficient heat is generated to ignite the fuel charge. This is shown by a puff of white exhaust smoke.

In the United States the method most commonly used utilizes a single cylinder engine of variable compression ratio. The engine operates with a fixed delay period of 13°, measured in terms of crankshaft angle before the top dead centre position of the piston. Thus the percentage of cetane in a blend requiring the same compression ratio for a 13° ignition lag as that required for the fuel under test gives the Cetane Number of this fuel (A.S.T.M. D 613–59 T).[3]

Diesel index.—This gives an estimation of ignition quality based on the aniline point and the specific gravity of a fuel.

$$\text{Diesel Index} = \frac{\text{Aniline Point °F} \times \text{API Gravity}}{100}$$

This formula determines the paraffinicity of a fuel, and since paraffins ignite more readily than the other hydrocarbons which are present it gives an indication of its ignition characteristics. It should only be used for petroleum fuels and then if no additives are present. Diesel Indices are of the order of three numbers higher than the corresponding Cetane Number for fuels with Cetane Numbers of about 45 to 50. This difference can vary widely however depending on Cetane Number level, fuel composition and crude oil source, and in some cases the Cetane Number is higher than the Diesel Index.

Significance of ignition quality.—The ignition requirements of a diesel fuel depend on the size and operating conditions of the engine in which it will be used. In general high ignition quality fuels give smooth combustion but in any engine an increased value above the minimum necessary for good combustion will not materially affect engine performance. Small high speed engines usually require fuels with ratings in excess of 40, but the slower speed engines with longer times available for combustion can utilize fuels with lower values. Fuels with high Cetane ratings enable engines to start more readily at low temperatures and result in engines achieving steady running conditions more quickly. Physical factors such as high air and fuel temperatures reduce ignition delay since these increase the rate at which the hydrocarbons attain ignition temperature. Since it has been shown that normal paraffins are more readily oxidized than olefines and aromatics[4] the chemical composition of a fuel is also important.

Additives.—The ignition quality of diesel fuels can be improved by using various additives[4] which promote the oxidation mechanism of the fuels. Some of the chemical types are:

(a) Alkyl nitrates.
(b) Aldehydes, ketones, ethers, esters and alcohols.
(c) Peroxides.
(d) Aromatic nitro compounds.

Carbon Residue

The carbon forming tendency is one of the most important properties of a diesel fuel and is determined by a carbon residue test. For high speed engine fuels, the carbon residue should be below 0·05%, and for medium speed engine fuels it should not exceed 1·5%. Fuels with carbon residues of up to 12% have been used without difficulty in some low speed engines which have adequate nozzle cooling and control of fuel viscosity. The nature of the carbon deposits is as important as the amount, since soft, fluffy deposits tend to be removed during the exhaust stroke whereas hard, tenacious deposits remain in the cylinder and cause poor combustion and cylinder wear.

Ash Content

Distillate fuels have low ash contents (below 0·01%) but in residual fuels ash contents may vary up to 0·1%. The ash consists of soluble and insoluble materials, and whereas the latter may be removed by efficient centrifuging, the former pass into the engine cylinders and are the main cause of cylinder liner and piston wear. This factor becomes increasingly important as engine speeds are increased.

Sulphur Content

The sulphur content of diesel fuels varies considerably depending on crude oil source. In general the sulphur contents of distillate fuels are below 0·1% but residual fuels may contain up to 4%. Some of the sulphur compounds present in distillate fuels may be removed by refining techniques such as Hydrofining, but it is not an economical proposition to reduce the sulphur content of the comparatively cheap residual fuels.

It is considered that the presence of sulphur compounds can cause cylinder corrosion, particularly in high speed engines. In addition these compounds can combine with carbonaceous materials from the fuel and lubricating oils to form hard, abrasive deposits. In some cases lubricating oils for diesel engines contain additives which neutralize the corrosive sulphur compounds formed during fuel combustion.

Specifications for Diesel Fuels

Standard specifications for Diesel Fuels have been issued by the British Standards Institution, B.S. 2869 : 1957,[5] the Class A and Class B fuels approximating to high speed and medium speed engine fuels respectively. These specifications are given in Table 22.

Diesel Fuels from Sources other than Petroleum

Diesel fuels produced from coal tar are chiefly aromatic in nature and because of their low cetane values, low calorific values, and high ignition temperatures, are not such good fuels as the corresponding petroleum oils. The removal of tar acids results in an improvement in quality, but the use of these fuels is restricted

TABLE 22
BSI Diesel Fuel Specifications

Class		A	B
Viscosity, kinematic, at 100°F, centistokes min		1·6 (a)	—
	max	7·5 (b)	14 (c)
Cetane number	min	45 (d)	— (e)
Carbon residue, Conradson	% wt max	0·1	1·5
Distillation, recovery at 357°C	% wt min	90	—
Flash point, °F	min	130	150
Water content	% wt max	0·1	0·25
Sediment	% wt max	0·01	0·05
Ash	% wt max	0·01	0·02
Sulphur content	% wt max	1·3	1·8
Sulphur, corrosive		Not more than slight tarnish	—
Strong acid number		Nil	Nil

(a) Approximately 30 Redwood 1 sec at 100°F
(b) Approximately 45 Redwood 1 sec at 100°F
(c) Approximately 65 Redwood 1 sec at 100°F
(d) A Diesel Index of 48 is normally sufficient to ensure a minimum Cetane Number of 45.
(e) The reproducibility of Cetane Number determination in the range likely to be encountered with Class B fuels is not yet entirely satisfactory. However, a fuel complying with the test requirements of Class B is unlikely to have a Cetane Number, or Diesel Index, below 23.

to low speed engines or to blending with high quality petroleum products. Blending in this manner results in the formation of a sludge, and the fuel must be desludged before use.

Diesel fuels produced from crude shale oil have similar properties to those obtained from crude petroleum.

High quality diesel fuels can be obtained by the Fischer-Tropsch synthesis, and these could be used for upgrading materials of poor quality, as shown in Table 23.[6]

TABLE 23
Characteristics of a Fischer-Tropsch Diesel Fuel

Diesel fuel	Fischer-Tropsch	Petroleum	50/50 Blend
Gravity, °API	51·1	35·1	42·7
Distillation, °C			
IBP	221	211	214
75% distils at	293	283	284
Pour point, °F	36	−9	19
Cetane number	90	50	69

Applications of Diesel Fuels

There is a growing preference for diesel-engined commercial vehicles in the UK, and at present about one quarter of all commercial vehicles are this type. These use about 2·5 million tons per annum of light diesel fuel, or DERV (diesel-engined road vehicles). The quantity of other gas/diesel oils produced is about 3·5 million tons per annum, and engines using these fuels find application in agricultural power units, in marine craft, for rail traction, and in various types of stationary power units. The use of diesel engines for rail traction has increased rapidly in the past few years, and will continue.

References

1. A. W. JUDGE, *High Speed Diesel Engines*, Chapman and Hall, London (1957)
2. *I.P. Standards for Petroleum and its Products*, Part II, Methods for Rating Fuels—Engine Tests, October (1960)
3. *A.S.T.M. Manual for Rating Diesel Fuels by the Cetane Method* (1959)
4. Improvement in Diesel Fuels, *Science of Petroleum*, Vol. 5, Part 3, p. 142, Oxford University Press (1955)
5. *British Standards Institution Publication*, B.S. 2869 : 1957
6. Data from Beacon Laboratories of the Texas Company, published in *The Science of Petroleum*, Vol. 5, Part 3, p. 147, Oxford University Press (1955)

CHAPTER 6

Atomization and Combustion of Fuel Oisl

IN ORDER to obtain good combustion it is necessary to produce suitable conditions for the hydrocarbons in the fuel to combine with the oxygen in the combustion air to form carbon dioxide and water vapour. It is the function of the burner to vaporize the oil or to disintegrate it into droplets, and to disperse the droplets into the combustion chamber in a manner such that vaporization and intimate mixing of fuel and air are achieved. To obtain the maximum thermal efficiency, the operation of the burner should result in complete burning of the fuel within the combustion chamber and with the minimum of excess air. Preheated air facilitates vaporization of the fuel and increases the flame temperature. Furthermore, the stack loss is smaller when waste heat recovery is used to preheat the combustion air.

Combustion of a liquid fuel takes place in stages, and the factors controlling these stages can be adjusted to regulate the rate of combustion (in certain applications very rapid heat release is required, whereas for some industrial processes slower combustion with retained flame emissivity is necessary). The stages are atomization, vaporization, mixing of fuel and oxygen, and ignition and maintenance of combustion, or flame stabilization.

An atomizing burner creates a liquid film which forms threads and finally droplets, these being dispersed in a controlled pattern into the airstream. The principal liquid properties which affect the degree of atomization are surface tension and viscosity, but since for all fuel oils under working conditions the variation in surface tension is small, then viscosity is of major importance. Optimum performance requires the correct viscosity for the particular type of burner being employed, and this viscosity is obtained by preheating the oil before it reaches the burner. The spray produced consists of droplets varying in size, which for good combustion should be within the range 0·03 to 0·015 mm diameter.

The droplets, on entering the combustion chamber, are heated by transfer from the flame, surrounding brickwork, and recirculated hot gases. The lighter fractions distil off and, if intimate mixing with air has been achieved and the temperature is sufficiently high for ignition, combustion of the vapour takes place in a reaction zone surrounding the liquid droplet. The heat released causes heavier fractions to be distilled, and these in turn are ignited and burn. Some of the vapours distilled from the liquid droplets are subjected to cracking and produce simpler hydrocarbons, hydrogen, and carbon. With residual fuels, cracking of the oil in the droplet stage may also occur, to give larger carbon particles which contain ash. The carbon produced from gas and liquid phase cracking is responsible for the luminosity of an oil flame, and under the required conditions of temperature and oxygen supply this carbon can be completely burned. When the carbon consumption is incomplete the particles pass out of the combustion zone to the stack, and when the concentration of stack solids is sufficiently high to be visible to the naked eye they are termed "smoke". In addition to its effect on the rate of combustion of the fuel vapour, the intimacy and rapidity of air and fuel mixing affects the rate of evaporation of a droplet.

Flame stabilization, or the maintenance of combustion, requires that the vapour/air mixture must be raised in temperature to the level at which the reactions persist. Generally, radiation from the surroundings and some form of recirculation of burned gases is used to ensure that sufficient heat is available for the fuel/air mixture to be brought to this required temperature.

The shape of the flame is determined by the direction of the spray of droplets issuing from the atomizer, by the direction and quantity of secondary air, and by the shape of the combustion chamber. Oil flames are adaptable and can be modified to suit the heat transfer requirements of a wide range of industrial processes.

The chemical calculations associated with the combustion of hydrocarbons, stack losses, and other necessary quantitative data are adequately dealt with in "The Efficient Use of Fuel".[1]

Oil Burners

An oil burner is required to deliver the fuel into the combustion chamber in a form suitable for combustion. This is achieved by vaporizing or "atomizing" the fuel, the former class of burners being described in Chapter 9.

Atomizing burners can be classified according to the source of energy used to disintegrate the oil. There are three types, namely:

  (i) pressure atomizers (pressure jets), in which pressure energy is employed,
  (ii) rotary atomizers, in which centrifugal energy is imparted to the fuel,
  (iii) twin fluid or blast atomizers, in which a gas is made to impinge on the liquid.

Pressure Atomizers

Oil, preheated to reduce its viscosity to 70 to 100 Redwood 1 seconds, is pumped at a pressure between 100 and 500 psig through a nozzle in order to produce the spray. The nozzle converts the pressure energy of the oil to kinetic energy, so that the oil is dispersed from the nozzle as a spray of very small droplets.

The oil enters the swirl chamber B (Fig. 13) through tangential ports A, rotates in the chamber around an air core, and issues through the orifice C in the form of a hollow conical film. The core expands due to centrifugal force, the film gets thinner, and finally disintegrates into droplets. The size and size distribution of the droplets is governed by the design and mechanical finish of the nozzle, the physical properties of the oil, and the pressure. For small nozzles the viscosity should be about 70 Redwood 1 seconds and for the larger nozzles should not exceed 100 Redwood 1 seconds.

The rate of discharge from a pressure atomizer varies as $\sqrt{P}$ and, in contrast with the case of a plain orifice, the rate of discharge decreases with decrease in viscosity. The droplet size increases with viscosity and varies inversely as $\sqrt[3]{P}$. The oil pressure can generally be varied between 100 and 500 psig so that, for a simple pressure atomizer, the discharge rate at the higher pressure is approximately double that at the lower pressure. The minimum pumping pressure is limited by the increase in droplet

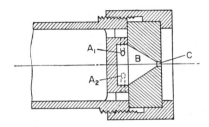

FIG. 13. Pressure jet nozzle.

size as the pressure is decreased. The turndown ratio of a burner is the ratio of the maximum to the minimum load, and in applications where the limited ratio of a pressure jet is a disadvantage, individual burners can be turned on or off, or "wide range pressure jets" can be used. These atomizers are claimed to operate with a good degree of atomization over a turndown ratio of up to 10 : 1 and are of several different designs. In the "spill" type (Fig. 14a) the velocity of oil through the tangential ports is maintained at low discharge rates by recirculating a controllable amount of the oil which enters the swirl chamber back to the suction of the pump. The return can be taken from the axis of the swirl chamber or from the periphery. Other designs vary the tangential port area (Fig. 14b) by the use of a sliding piston which controls the number of ports uncovered.

The air for combustion is admitted through an air register which surrounds the burner. The air register controls the amount of air, the mixing of oil and air, and also imparts a rotary motion to the air which creates a zone of low pressure along the axis of the air register and so brings about recirculation of the burning gases to help stabilize the flame. The spray being emitted from the nozzle thus receives heat by radiation from the flame and surroundings, and also from contact with the recirculated hot gases.

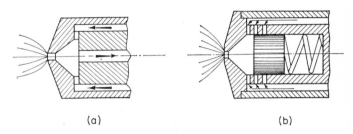

Fig. 14. "Wide range" pressure jet nozzles.

Pressure jet burners are mainly used for firing large steam boilers, furnaces, and hot water boilers, since they require a smaller amount of energy than any other type of atomizer. For simple, or fixed pressure jets the limited turndown ratio is a disadvantage, and where automatic control is required it is best to use wide range pressure jets. For good performance accurate machining is vital, accurate fuel temperature has to be maintained, and for small sizes great care has to be taken to filter the fuel.

Rotary Atomizers

Atomization of the oil is achieved by centrifugal force, the oil flowing through a central pipe to the inner surface of a revolving hollow tapered cup. The cup is rotated at high speed by means of an electric motor or by an air turbine driven by a proportion of the atomizing air. Friction between the oil and the wall of the cup causes the oil to rotate with the cup, and centrifugal force, together with the taper of the cup, causes it to be discharged from the rim at a high velocity. Owing to the uniformity of the surface of the cup the thickness of the oil film and the velocity of the oil are the same over the circumference of the rim of the cup. Thus good atomization is obtained and the size range of the droplets is closer than obtained with any other type of atomizer. About 15% of the air required for combustion is supplied at 4 to 10 in. wg as primary air around the outside of the cup. This air supply is necessary to prevent the oil droplets being dispersed from the rim of the cup along a plane at right angles to the axis of the cup, and also to exert additional atomizing forces on the oil droplets. Figure 15 shows diagrammatically the arrangement of an air driven rotary cup atomizer. For motor driven cups it is common to have the atomizing fan, the cup, the pump, and motor mounted as a single unit. The viscosity of the fuel should be 100–300 Redwood 1 seconds and the pressure should be low, about 2–3 psig. The rotary atomizer is much less sensitive to viscosity than the pressure jet and is much less liable to chokage by grit. On the other hand, this type of burner is more susceptible to deposition of carbon, caused by radiation from the surroundings after the burner is shut down.

Twin-fluid or Blast Atomizers[2]

Blast atomizers are of three main types:

(i) low pressure, utilizing air at $\frac{1}{2}$ to 1 psig as the atomizing medium,
(ii) medium pressure, utilizing air at 5 to 15 psig,
(iii) high pressure, utilizing air, but more frequently steam, at a pressure exceeding 15 psig.

These types are further subdivided into "inside mix" and "outside mix". With the former the steam or air and oil impinge and mix within the burner and issue as a foam or fog, depending on whether the atomizing fluid is passed into the oil or the oil sprayed into the atomizing fluid. In the "outside mix" type the oil is released into the atomizing fluid at the outlet from the burner, and the air or steam is given a prerotational swirl before it meets the oil.

The atomization of the fuel requires approximately the same kinetic energy on burners of the three pressure ranges, so that

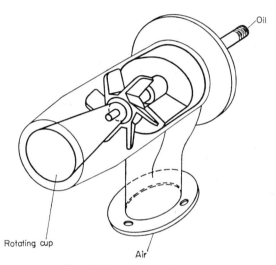

Fig. 15. Rotating cup burner.

the higher the pressure the smaller the quantity of air or steam required. Well designed high pressure burners require about 0·3 lb of air or steam per pound of oil, medium pressure burners about 0·7 lb of air per pound of oil (4% of total air for combustion) and low pressure burners about 20 to 100% of the combustion air. The temperature of the atomizing medium should not exceed 400°F in order to avoid cracking of the fuel within the burner, and where regeneration or recuperation is used to provide preheat to as much of the combustion air as possible, high or medium pressure burners should be used.

Low pressure burners.—Since a large proportion of, and sometimes all the combustion air is passed through the burner, a turndown ratio of only 2 : 1 to 5 : 1 is possible without adversely affecting the atomizing efficiency. In cases where such a low turndown can be accepted the LP burner is probably the most economical, efficient, and reliable. The air is supplied by means of a single stage centrifugal fan, which is relatively simple, cheap, and easy to maintain. When all the combustion air is used for atomization, good air/oil mixing is achieved and rapid combustion with the minimum of excess air results. The turndown ratio in this case does not normally exceed 2 : 1, compared with 5 : 1 when about 20% of the combustion air is used for atomization. When a negative pressure exists in the combustion chamber, burners with a 5 : 1 turndown ratio can be used, the remainder of the air (secondary air) being induced around the burner through regulators which are designed to produce good air/oil mixing. In furnaces which use balanced or slight positive pressure a separate fan must be used to provide the secondary air at a pressure of a few inches wg, and this is only justified if LP burners are used in conjunction with highly preheated secondary air or in furnaces with a large heat release where an appreciable saving in power is obtained by providing the bulk of the air at a lower pressure than that required for atomizing.

Figure 16 shows an LP burner in which only a proportion of the combustion air is used for atomizing. The secondary air can then be adjusted to give effective control of furnace temperature and/or furnace atmosphere. For best results the oil should have a viscosity at the burner of not greater than 70 Redwood 1 seconds. LP burners are used for metallurgical and other furnaces, in applications requiring fuel outputs as small as 2 gallons per hour and less, and for small and medium boiler plant where their greater turndown ratio and smaller flame are advantages over fixed pressure jets.

Medium pressure burners.—Whereas in LP burners at least 20% of the combustion air is required for atomization and the turndown ratio is consequently limited to 5 : 1, with MP burners at maximum output, less than 10% of the combustion air is required for atomization. Consequently a turndown ratio of up to 10 : 1 can be obtained without detriment to the atomization and combustion efficiency, the secondary air being separately controlled. Furthermore, since the majority of the air is used as secondary

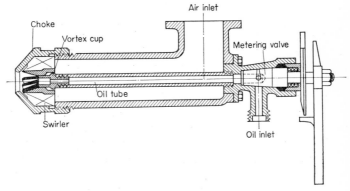

FIG. 16. Low pressure air burner.

air, more heat recovery can be used than when operating LP burners.

The air is supplied by a rotary compressor at pressures between 5 and 15 psig, and although the power requirement is frequently less than required by an equivalent LP burner, the compressor is more expensive than the LP fan.

MP burners are used in boiler and furnace work where a good range of control is required. They are very suitable for use in high temperature regenerative or recuperative furnaces because the small amount of primary atomizing air required permits about 95% of the total air to be preheated.

High pressure burners.—HP air burners are very similar to MP air burners but require a more expensive compressor. In very few cases, other than where a compressed air supply is available, is it necessary to use HP rather than MP air burners. Steam is more frequently the atomizing fluid used in HP burners, and these find

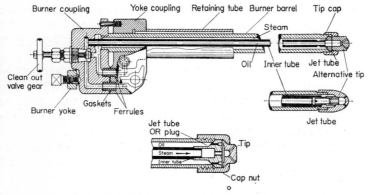

Fig. 17. Steam atomizing burner.

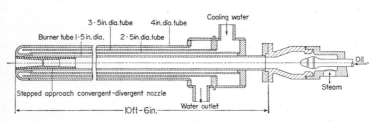

Fig. 18. Steam atomizing burner suitable for open hearth furnaces.

applications in the open hearth steel furnace, in boilers, since no fan or compressor is required, or in cases where a plentiful supply of steam is available.

Figure 17 shows a steam atomizing oil burner assembly for use in most applications and Fig. 18 a typical water cooled burner for use in open hearth furnaces. The turndown ratio for steam atomizers is up to 10 : 1.

When used for firing boilers these burners use about 0·3 lb steam per pound of oil, or approximately $2\frac{1}{2}\%$ of the steam generated. Their disadvantage in this application is that separate equipment has to be used for raising atomizing steam at start-up. They are most useful for firing boilers where no source of power is available, and in such cases a small auxiliary steam boiler fired by a hand operated kerosine burner is used as lighting-up equipment.

Steam atomizers of the "internal mix" type are widely used in open hearth furnaces, where the main function of the steam is to entrain the air required for combustion and to control the flame length and combustion rate. The determining factor in these large high temperature furnaces is not oil droplet size (which is small enough to have little effect on the rate of combustion) but the rate of mixing of oil and air. Steam is used as the atomizing medium in preference to air since most open hearth furnaces have a waste heat boiler fitted. The work of the International Flame Research Foundation[3] has shown that, whereas the temperature of an air-atomized flame is higher, the total radiation from a steam-atomized flame is as high as that from a flame produced by air atomization, and in the latter case the radiation is more uniformly distributed over the flame length.

Apart from the open hearth furnace, steam atomizers are rarely used in metallurgical furnaces owing to the lower flame temperature compared with air atomizers, and also since the extra water vapour can produce higher rates of scaling of the furnace stock and results in a greater stack loss.

Efficient Combustion

One of the most difficult conditions to achieve in burning liquid fuels is the intimate mixing of the fuel and air in order that combustion can be completed with the minimum of excess air and hence with maximum thermal efficiency. As the excess air is reduced, a stage is reached where some of the fuel remains unburned, since sufficient air is not available to react with all the fuel in the limited time available and under the conditions prevailing in the combustion chamber. The unburned fuel takes the form of carbon particles and the combustible gases carbon monoxide and hydrogen. The carbon particles produced by the cracking of the vapours obtained from the fuel, together with the carbon-plus-ash particles derived from the cracking of the liquid droplets of residual fuels, constitute the "stack solids". As previously stated, when the concentration of these solids is high enough to be visible they are called "smoke", but at all concentrations they constitute a loss of unburned fuel and a source of atmospheric pollution.

The quantity of stack solids can be measured by the following methods:

(a) Filtration of the stack gases through a silica wool or similar siliceous filter at a controlled rate, and weighing the filter plus deposits.

(b) Filtration of a measured quantity of the stack gases through a filter paper and matching the discolouration against a scale. This method is the basis of the Shell Smoke Number Test,[4] in which the shades are numbered from 0 (white) to 9 (black). Visible smoke corresponds to Shell Smoke Number 6. No unique correlation exists between the weight of stack solids and the Shell Smoke Number, since the method is sensitive to the particle size distribution of the stack solids. In some installations the method does provide a good qualitative guide to indicate improvements in combustion conditions.

(c) Measurement by means of a photo-electric cell of the reduction in light intensity across a fixed thickness of gas caused by the stack solids. The results obtained by this method also fail to provide a unique correlation with the quantity of stack solids, since the reduction in light intensity is sensitive to particle size distribution.

Combustion efficiency can be assessed from a measurement of the stack solids and the flue gas analysis. The stack solids figure permits the quantity of unburned solid fuel to be calculated, and the flue gas analysis measures the unburned gases. Hydrogen is not usually determined, but a good approximation is $H_2 = 0.5CO$.

Table 24 shows the results obtained in a boiler plant trial,[4] and it can be seen that as the excess air is reduced from 59% to 3% the increase in stack solids is from 0.24% to 0.39%, the increase in CO content from 0 to 1.0%, and visible smoke (Shell Smoke No. > 6) is produced. In this particular case where the combustion conditions are first class, the decreased stack loss achieved by reducing the excess air outweighs the increased stack loss caused by increased stack solids and carbon monoxide. It must be emphasized that the relationship of stack solids and carbon monoxide content of flue gas to excess air shown in Table 24 applies only to the boiler under test, and that in order to achieve the highest combustion efficiency in any particular installation its performance should be studied in the same way. However, combustion efficiency in itself might not be the sole objective, and consideration must be given to the fouling of heating surfaces by carbonaceous material and to the problem of atmospheric pollution.

TABLE 24

Tests Carried Out on an 8000 *lb/hr Economic Boiler at an Output of* 4800 *lb/hr*

Flue gas analysis, % vol.			Excess air %	Shell smoke number	Stack solids, % fuel, wt.
CO_2	O_2	CO			
9.7	8.0	0	59	3	0.24
12.4	4.5	0	26	4	0.28
13.4	2.7	0.5	12	5	0.31
14.2	1.3	1.0	3	7	0.39

Low Temperature Corrosion

The subject of low temperature corrosion is dealt with in this chapter since in order to achieve maximum thermal efficiency it is necessary to recover the maximum amount of heat from the flue gases, and it is in the pursuit of minimum stack losses that troubles due to acid condensation and corrosion are most often encountered.

One of the major problems associated with the use of solid and liquid commercial fuels of high sulphur content is the tendency of the flue gas to cause corrosion and deposits on relatively cool surfaces. Although this is particularly the case in high efficiency boilers, low temperature corrosion has been experienced in a wide range of fuel burning equipment where adequate precautions have not been taken. The corrosion of low temperature surfaces, the formation of deposits, and the emission of smuts from stacks are caused by the condensation of sulphuric acid from the flue gases. The acid results from the combination with water vapour of traces of sulphur trioxide which has been formed by the oxidation of sulphur dioxide produced in the combustion system. Sulphur trioxide occurs in flue gases in concentrations of 0·0005–0·005% and causes the dewpoint to rise from the water dewpoint of 100–120°F to temperatures as high as 350°F. Thus the heat recovery from the flue gas, and hence the efficiency, is limited by the acid dewpoint. The amount of condensation appears to depend not so much on the temperature of the flue gases as on that of the heating surfaces. The corrosion of air heaters and other cool parts of a boiler can be very severe, and an instance has been reported[5] where a new boiler had to be withdrawn from service after less than six weeks in operation.

Instruments and techniques have been developed[6] to measure SO_3 concentration, dewpoint temperatures, the rate of condensation of acid, and the rate of corrosion of a cooled metallic specimen. Whereas the dewpoint temperature gives the range over which corrosion will occur, the rate of acid build-up determines the severity of corrosion. This rate is maximum at about 50°F below the dewpoint, so that a typical dewpoint of 260°F would give rise to severe corrosion at about 210°F.

A relationship between sulphur content of the fuel oil and the acid dewpoint is shown in Fig. 19,[7] from which it can be seen that there is a rapid increase in the dewpoint up to a sulphur content of 1%, but that from 1 to 5% sulphur the increase in

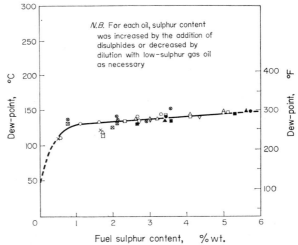

FIG. 19. Relationship between acid dewpoint and sulphur content of fuel.

dewpoint is very gradual. The removal of sulphur from normal fuel oils in order to combat low temperature corrosion problems only would be uneconomic at the present time, so that investigational work has been directed towards inhibiting the formation of sulphuric acid and also towards reducing the effects of condensed acid. These methods include alterations to the combustion system in order to modify the flue gas composition, the use of additives to the fuel or as injections into the flame or flue gas system, and the protection of metallic surfaces by additives. Heating surfaces should always be maintained in a clean condition, since dirty surfaces aggravate corrosion troubles, possibly owing to catalysis of the trioxide formation by the deposits.[8]

Minimum Excess Air

It has been shown[5] that operating with the minimum of excess air reduces low temperature corrosion by as much as a factor of 30 because of the reduction in the formation of sulphur trioxide. In cases where the reduction in excess air affects the heat transfer in a boiler by reducing the convective heat exchange, this can be restored by recirculating hot flue gases from the economizer outlet back to the furnace, thereby increasing the mass velocity of the gases in the boiler.

The Use of Additives

Dolomite and magnesium carbonate have been used,[9] but although the SO_3 content of the flue gas can be reduced from 15 ppm to about 4 ppm, and the dewpoint from 300°F to 240°F, superheater blockage by the products of reaction can be extensive.

Zinc dust injection[10, 11] gives a smoke of zinc oxide which is effective in reducing the dewpoint to as low as 120°F and in reducing the deposition of sulphuric acid, although fouling of superheaters occurs but at a lower rate than with dolomite.

Ammonia gas[7, 12] can be used with success as it reacts selectively with the sulphur trioxide in the flue gases when injected in the optimum temperature range and at the correct concentration to form neutral ammonium sulphate, the sulphur dioxide in the flue gases remaining uncombined. The correct position for the injection is at the highest practicable point upstream where the boiler metal temperatures do not exceed 420°F. Metal temperatures in the range 420 to 520°F must be avoided, otherwise some molten ammonium bisulphate may be formed, which can cause a build-up of solid deposits and may lead to blockage. The optimum concentration is only slightly more than stoichiometric concentration for the reaction and is normally within the range 0·05 to 0·10% by weight on the fuel, depending on the sulphur trioxide content of the flue gases. Most of the products of the reaction are carried away in the flue gas stream, but some are deposited on the boiler tubes and are easily removed by periodic water washing. Provided that ammonia is injected in the correct position and at the correct concentration successful results are obtained with both oil and coal fired boilers. Reductions in the corrosion of 75 to 80% are usual and the acid dewpoint is completely eliminated.

Heterocyclic tertiary amines obtained from coal tar have been injected into the flue gases[13] where they evaporate and subsequently condense on the cooler metal surfaces as a liquid film which resists attack by sulphuric acid. This method has shown success in reducing the attack on air heater surfaces, but unless great care is taken to control the quantity of additive, rapid blockage can occur due to excess tar base.

The literature on the subject of low temperature corrosion is extensive and work continues. It would appear that many of the difficulties encountered with the injection and use of some additives are as yet unresolved and that the most promising method of reducing corrosion is to operate with the lowest possible excess air and to maintain the low temperature surfaces above the acid dewpoint.

Smut Formation and Steel Stack Corrosion

The emission of smuts results from a combination of the circumstances which produce stack solids together with those causing acid deposition in a steel stack. The carbonaceous material adheres to the sulphate of iron formed by acid attack on the steel and, in conditions of change of temperature or gas flow, flakes are formed and entrained out of the stack.

A successful method [14] of overcoming this trouble of corrosion and smut emission is to fit an aluminium shield around the stack to form a ¼ in. air gap, thus insulating the stack and maintaining the metal temperature above the acid dewpoint.

References

1. *The Efficient Use of Fuel*, 2nd edition, H.M.S.O., 1960
2. R. P. FRASER, *Proc. Joint Conf. on Combustion*, 1955, Inst. Mech. Engrs.
3. E. H. HUBBARD, *J. Inst. Fuel*, **32**, 328 (1959)
4. K. H. SAMBROOK, *Petroleum*, **17**, 56 (1954)
5. B. LEES, *J. Inst. Fuel*, **29**, 171 (1956)
6. *Testing Techniques for Determining the Corrosive and Fouling Tendencies of Boiler Flue Gases*, The Boiler Availability Committee (1961)
7. L. K. RENDLE and R. WILSDEN, *J. Inst. Fuel*, **29**, 171 (1956)
8. H. E. CROSSLEY, *Proc. Conf. Liquid Fuel Firing*, Inst. Fuel (1959)
9. T. J. WILKINSON and D. G. CLARKE, *J. Inst. Fuel*, **32**, 61 (1959)
10. G. WHITTINGHAM, *J. Soc. Chem. Ind.*, **67**, 411 (1948)
11. P. A. ALEXANDER *et al.*, *J. Inst. Fuel*, **34**, 53 (1961)
12. L. K. RENDLE *et al.*, Fire side corrosion in oil-fired boilers, *5th World Petroleum Congress* (1959)
13. E. BRETT-DAVIES and B. J. ALEXANDER, *J. Inst. Fuel*, **33**, 163 (1960)
14. H. A. BLUM, B. LEES and L. K. RENDLE, *J. Inst. Fuel*, **32**, 165 (1959)

CHAPTER 7

The Use of Liquid Fuels in Boilers, Industrial Furnaces, and Gas Turbines

THE advantages of liquid fuels make them particularly attractive for many types of industrial furnaces and boilers, and the rate of increase in consumption of petroleum fuels from 1938 to 1960 is shown in Fig. 20.[1]

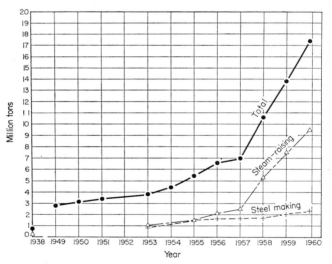

FIG. 20. Fuel oil consumption. U.K., 1938–1960.

The principle advantages of liquid fuels are:
(a) Negligible variation in quality.
(b) Cleanliness and absence of ash.

(c) Ease of handling and storage, resulting in saving in labour.
(d) Good control and flexibility, resulting in high furnace efficiencies.
(e) High flame temperatures and luminous flames.
(f) High rates of heat release, resulting in higher output rates from furnaces.

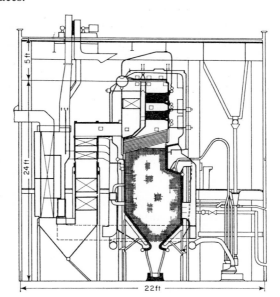

Fig. 21. (Left) Oil-fired marine boiler 240,000 lb/hr evaporation. (Right) Pulverized-fuel-fired land boiler. 240,000 lb/hr evaporation. The figure illustrates the relative difference in size of the two boilers required for the same duty.

The air requirements for a particular grade of oil are virtually constant, so that automatic control is facilitated and high furnace efficiencies should result. Liquid fuels give flames of high emissivity, and this property, together with higher flame temperatures than obtained with most industrial gaseous fuels, result in high rates of heat transfer. In addition, the ash content of petroleum fuels rarely exceeds 0·1% (which is of the order of 1% of the ash content of most industrial coals). The low ash content and the high heat release permit the furnace of an oil fired boiler to be rated much higher than a coal fired boiler, and this effect is vividly illustrated in Fig. 21.[2] The rate of heat release in a pulverized coal fired boiler is about 30,000 B.T.U./(ft^3) (hr), whereas in a marine oil fired boiler the heat release can be as high as 1,000,000 B.T.U./(ft^3) (hr). For power stations and industrial plant, boiler availability has to be considered and the designs are normally for heat release rates which do not exceed 100,000 B.T.U./(ft^3) (hr).

The technical advantages of oil firing, together with the progressively more competitive prices for petroleum fuels have continued to encourage the consumption of these fuels in industry. Whereas in 1938 fuel oils (excluding gas oils) represented 7% of the total petroleum products consumed in the UK, in 1960 the proportion was slightly greater than 40%.

Boilers

As can be seen from Fig. 20 the use of liquid fuels for steam raising has increased ten-fold in the period 1953 to 1960, mainly on account of the Central Electricity Generating Board's decision to convert certain of its plant from coal to oil firing and to construct plant specifically for alternative oil or coal firing.

Large Water Tube Boilers

The typical oil used in these installations is a residual fuel of viscosity between 1500 and 3500 Redwood 1 seconds at 100°F and having a sulphur content of approximately 4%. An installation includes heated tanks with steam traced and lagged pipework, pumps and other items of equipment, the steam in large installations being provided by boilers separate from the main plant. Viscosity controllers can be used to regulate automatically the viscosity of the oil delivered to the burners by controlling the heat input to the oil heaters, but their cost is high, installation is difficult, and there exists some doubt as to whether they are necessary.

The problems associated with low temperature corrosion of air heaters and other low temperature sections of a boiler have been discussed in Chapter 6, and to date hardly any trouble has been experienced with vanadium attack on high temperature uncooled metal surfaces.

Most of the large oil burning installations are conversions of plant originally designed for pulverized coal, most of which use pressure atomizing burners in preference to steam atomization, the latter having a steam consumption of about 1% of the total evaporation. Steam atomizers have been used in one installation[3] for start-up conditions, when the superior atomization reduces combustion losses in a cold furnace, pressure burners then being used when the furnace reaches normal operating temperature. Automatic ignition, flame failure devices, and remote controls are incorporated in a few modern burner systems.

One of the greatest difficulties encountered in converted boilers is in attaining the required steam temperature. Oil flames radiate more than pulverized coal flames, require less excess air, and maintain clean furnace wall tubes. Consequently, since the radiant heat transfer in the steam generator section is greater, and since the temperature and mass flow of gases passing to the superheater (where the heat transfer is by convection) is lower, the final steam temperature is less than with pulverized coal. One method of restoring the heat transfer to the superheater is by recirculating flue gases from the outlet of the economizer back to the furnace. In this way the greater flow of gases increases the rate of convective heat transfer in the superheater and tends to compensate for the reduction in gas temperature. An earlier alternative method of maintaining high gas flow rates was by the use of high excess air rates, but this was found to aggravate low temperature corrosion troubles and increase the stack loss. Other methods used to maintain the temperature of the gas flowing to the superheater include the installation of high level burners, and insulation of a proportion of the radiant tubes by modifying the baffle system.

A full scale experimental boiler has recently been built by the

British Petroleum Company,[4] and this will be used to study the whole range of problems, such as those arising from conversions of coal-fired boilers to oil firing, low temperature and high temperature deposits, corrosion, and air pollution.

Shell-type Boilers

Shell boilers of the Lancashire or Economic type, with ratings from 1500 to 16,000 lb/hr, have been converted or specifically purchased for oil burning. The fuels used include those having viscosities of 200, 950, 1500, and 3500 Redwood 1 seconds at 100°F, and whereas most burners were of the pressure jet type, rotary cup burners and low- or medium-pressure twin fluid burners are becoming more popular. One of the reasons for this trend has been the difficulty in obtaining satisfactory turndown with pressure jets.

The furnace tube wall of shell boilers, being water cooled, presents a large, relatively cool surface to the flame. In order to obtain complete combustion of the heavier grades of fuel oil in such boilers, it is sometimes necessary to line part of the furnace tube with refractory brickwork, as shown in Fig. 22,[5] the lining covering about 120° of the circumference. The lining radiates heat back to the flame and thus aids flame stability and rapid combustion.

With coal-fired Economic boilers the quantity of heat transferred in the furnace tubes is about equal to that transferred in the smoke tubes, whereas when the boiler is fired with oil a larger proportion of the heat is transferred in the furnace tubes. Consequently the temperature of the flue gases passing to the smoke tubes is reduced and the convection heat transfer impaired. In converted Lancashire boilers, which by design have a limited area for convection transfer, it is usually necessary to fit an economizer in order to achieve reasonable efficiency.

In recent years "packaged" boilers have been manufactured in the UK, and in these boilers, which are self contained, single, compact units, completely preassembled by the manufacturer, the radiant and convection heating surfaces are suitably designed for oil firing.[6] Boilers of this type (Fig. 23), which are capable of efficiencies of 80%, can be obtained for evaporations up to 25,000 lb/hr. Pressure atomizing, air atomizing, or rotary cup burners are used, and the boilers are fully instrumented with complete automatic controls. The latter include automatic modulating oil burner control, synchronized air control, flame failure

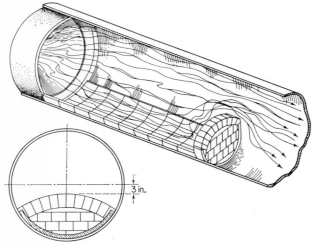

Fig. 22. Furnace tube lining.

device, and water level control. These boilers are highly rated and correct feedwater treatment is necessary to avoid scale formation and consequently overheating of metal surfaces.

Industrial Furnaces

Oil firing is used in a wide variety of industrial furnaces,[7, 8] those using the largest quantities of fuel oil being the open hearth steel furnaces and glass furnaces. A recent development in the steel industry is the use of heavy fuel oil in blast furnaces.

Fig. 23. Sectional diagram of a Steambloc boiler showing the three gas passes.

1. Three-pass boiler
2. Patent twin rear door
3. Oil Burner
4. Forced-draught fan
5. Modulating pressure control
6. Water level control
7. Boiler feed system
8. Partially corrugated flue
9. Boiler insulation
10. Control cabinet
11. Clean-out door
12. Lifting lugs
13. Steel bedplate
14. Alternative oil burner manual control

LIQUID FUELS IN BOILERS, FURNACES AND TURBINES 101

Furnace Temperature Control

Many processes require accurate control of furnace temperature in order that the desired temperature of the furnace stock can be achieved without overheating, or in order to vary the stock temperature on a time basis. To maintain a low furnace tempera-

FIG. 24. L.P. Air Burner suitable for low temperature furnaces.

ture (less than 850°C) with oil firing can be a difficult operation, since flame instability can result from too low a heat release from the burner.

Until recent years furnace temperature control was achieved either by use of "surplus" excess air or by the recirculation of flue gases, the latter method being preferable since it does not lead to an increased heat loss in the flue gases. Furthermore, the excess air method increases the oxygen content of the furnace atmosphere and could lead to excessive scaling of the furnace stock. In some cases where recirculation is used the jet action of the flame can supply the energy required, but most often a blower is used and it is usually necessary to maintain the temperature of the blower below 700°C.

Burners are now available which maintain stable combustion conditions with furnace temperatures as low as 650°C, or even lower if used in conjunction with flue gas recirculation. One such burner is shown in Fig. 24, a flame tunnel which incorporates a refractory recirculation device being used with a low pressure air burner through which all the combustion air passes. In this way high intensity combustion is achieved in the tunnel at low rates of fuel input, and stable combustion with accurate control can be maintained in furnaces which are operated at a low temperature. The O.C.C.R. gasifier (see Chapter 8) is another type of burner which enables fuel oil to be burned under stable conditions in low temperature furnaces. In this two-stage burner the oil is gasified by partial combustion with a proportion of the theoretical air in a pre-combustor, the product gas and suspended carbon then being burned to completion in the furnace.

Furnace Pressure Control

The control of furnace pressure has been widely used in open hearth furnaces and in glass tanks, and more attention is now being given to this aspect of control in heating and heat treatment furnaces.[9] A pressure measuring device, connected to a tapping which is placed in the roof of one of the walls, is linked to a regulator which controls the furnace damper. Correct pressure control reduces air infiltration and thus increases the efficiency of the furnace by decreasing the stack loss. In addition, the chilling of the furnace stock by impingement of cold air is avoided, and since all the air for combustion is supplied to the burners, control of excess air is simplified. When air infiltration has been eliminated by an increase in furnace pressure, any further increase in pressure results in a heat loss owing to the hot products of combustion being expelled from the furnace. Such emission of hot gases can cause excessive refractory wear and difficulties in operating the furnace owing to "sting-out" of the flames.

Combustion Control

The use of fuel/air ratio control is essential in any furnace in order to obtain maximum efficiency. An oxygen analyser or CO_2 recorder installed in the flue can be used to measure and record the composition of the waste gases, and in addition to adjust the fuel/air ratio control whenever necessary. Control of furnace pressure to prevent air infiltration is essential if fuel/air ratio control is to be effective.

Open Hearth Furnace

The open hearth method of steel manufacture is a batch process carried out in a regenerative furnace which is operated near the temperature limit (1680°C) of the refractories used in its construction. The original fuel was hot raw producer gas, but for basic open hearth furnaces this has been largely replaced by liquid fuels, since the latter give greater flexibility, higher flame radiation, easier control, higher rates of heat input to the furnace (hence higher output rates), and lower heat requirements per ton of steel. High pressure twin-fluid burners are used as the regenerators provide combustion air at a temperature of about 1100°C, steam being the usual atomizing medium because most furnaces are fitted with waste heat boilers.

The essential requirements of an open hearth furnace flame is high emissivity, and this property has been shown to be dependent on the C/H ratio of the fuel.[10] Thus pitch/creosote coal tar fuel (with a C/H ratio of about 14) is superior to the normal heavy petroleum fuel oil of C/H ratio about 7, but the addition of carbon black to petroleum fuels is used to increase the C/H ratio and hence the emissivity. Coal tar fuels have a sulphur content of 1% maximum, and this specification can be matched in petroleum fuels by blending from selected crudes. Such a specification is necessary for the production of low sulphur steels (0·025%S). The supply of coal tar fuel is limited, and over one and a half million tons of heavy petroleum fuel oil is now used each year in open hearth furnaces. A typical specification for a petroleum fuel used in the production of low sulphur steel is given in Table 25.

TABLE 25

Specification of Petroleum Fuel Oil Used in the Manufacture of Low Sulphur Steel

Flash point, °F	150 min
Specific gravity, 60/60°F	0·995 max
Sulphur content, % wt	1% max
Viscosity, Redwood 1 sec at 100°F	1500–3500
Viscosity cs. at 122°F	175–370
Ash, % wt	0·1 max

The rate of combustion in open hearth furnaces is governed by the rate of mixing of fuel and preheated air, and this in turn is governed by the momentum of the flame. The burners are required to atomize the fuel and provide the conditions necessary for correct flame length and direction. It has been shown[11] that the burner should be of the internal mix type with a convergent/divergent nozzle (Fig. 18), and the steam consumption is about 0·5 lb/lb oil. Much work has been carried out on the design of burners and the aerodynamics[12] and heat transfer[13, 14] in open hearth furnaces. The results, which have been applied to obtain accelerated combustion and more symmetrical flames, have led to considerable improvements in production and fuel consumption.

Recent developments include the use of oxygen for flame enrichment during the scrap melting stage of the process, and the use of oxygen via roof lances during the refining stage.

Glass Furnaces

The glass industry uses fuel oil for the firing of tank furnaces, pot furnaces, and annealing furnaces. The bulk of the glass is manufactured in tank furnaces, while special and more expensive glasses are melted in pot furnaces.

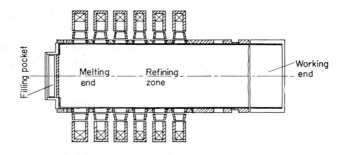

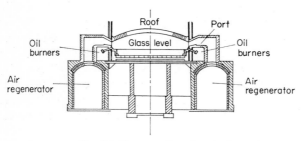

Fig. 25. Regeneration glass tank.

Glass tanks are regenerative and similar to open hearth furnaces, but work on a continuous basis, raw materials being fed at one end and the glass withdrawn at the other by forming machines. These tanks operate with a crown temperature of 1500 to 1550°C, and there are several firing ports at the melting end (Fig. 25).

Fuel oil has been steadily replacing producer gas for firing glass tanks and the grade most commonly used has a viscosity of 950

Redwood 1 seconds at 100°F. Since the surface of the glass is exposed to the flame, freedom from contamination of the charge by the inorganic constituents of the fuel oil is of major importance. Sulphur in the fuel can cause small blisters of sodium sulphate to be formed in the glass, can cause yellowish staining, and can lead to deposits of sodium sulphate on refractories which results in increased wear. A maximum sulphur content of 2·5% is considered as acceptable to the industry,[15] but for glass bottle manufacture the sulphur content can be as high as 3·5%. Vanadium in fuel oils also leads to discoloration of the glass, and a maximum of 150 ppm is desirable for special types of glass. Although coal tar fuels give flames of higher luminosity than petroleum fuels, this has not been held to have any significance in glass tanks,[16] although some controversy still exists on this point.

The burners used should give flames of high heat release, a flame length which does not exceed the width of the furnace, and the same requirements of jet momentum and air entrainment apply as for the open hearth furnace. Good atomization is necessary to prevent impingement of oil droplets on the surface of the glass. Pressure jet, medium and high pressure burners have been used, the latter being in more general favour. Various systems of burner installation have been used, the best being the underport type, in which the burner or burners are situated below the air port (Fig. 25). This method gives a flame which does not strike the glass and the burners do not require water cooling.

Blast Furnaces

Recent development work[17, 18] has shown that fuel oil can be used successfully to replace part of the coke charge to blast furnaces. The oil is injected into the hot blast tuyeres which are distributed around the periphery of the furnace at a level slightly above the hearth. A normal ring main distribution system (Fig. 26) supplies oil to the metering pumps, one such pump serving each tuyere. The oil is injected into the blast (Fig. 27) at a pressure two pounds higher than the blast pressure and at a temperature of

180°F, which gives a viscosity at the nozzle of about 200 Redwood 1 seconds.

The oil used is of up to 3500 Redwood 1 seconds at 100°F, with

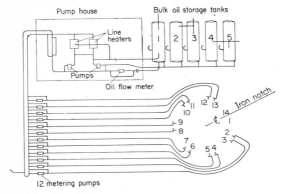

Fig. 26. Oil injection system for a blast-furnace.

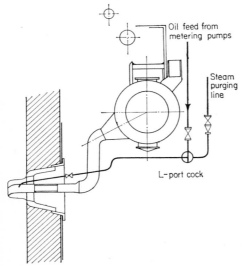

Fig. 27. Distribution of fuel oil into blast.

a sulphur content of up to 3·5%. Fifteen to eighteen gallons of oil can be used to replace about two hundredweights of coke per ton of iron produced, the output of the furnace is increased by about 10%, and there is a significant saving in the cost of iron production. Furthermore, fuel oil injection permits moderate expansion of iron production without the normal capital expenditure on increasing coke producing capacity. Similar development work is being carried out in America, Canada, Belgium, and Germany.

Gas Turbines

Gas turbines are finding increasing applications for power production and propulsion, the main reasons being their relative simplicity. Other advantages include high power/weight ratio, cheap and simplified lubrication, quick development of full power from start-up, low capital cost, and freedom from cooling water requirements.

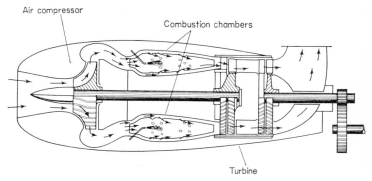

Fig. 28. Essential components of gas turbine.

A simple open circuit gas turbine consists of three essential components, (a) a centrifugal air compressor, (b) one or more combustion chambers in which fuel is burned continuously, and (c) a turbine revolving on a common shaft with the compressor

(Fig. 28). Air is induced into the compressor and forced at a pressure of about 60 psig into the combustion chamber, where it meets a continuous supply of fuel which is pumped at a pressure of about 800 psig. An electric motor initially spins the shaft and the fuel is ignited with a spark plug. The compressor supplies a quantity of air equivalent to an air/fuel ratio of about 60/1, the combustion air of about 14/1 being supplied to the combustion chamber (or flame tube) and the remainder is passed around the flame tube to cool the flame tube wall. This air is then passed through holes in the flame tube wall to mix with the hot combustion gases and so reduce their temperature from around 1800°C to a temperature which is dependent on the type of fuel. For distillate fuels the gases are cooled to about 900°C, whereas for residual fuels this temperature should not exceed 680°C in order to avoid troubles arising from the inorganic ash constituents of such fuels. This dilution of the combustion gases with relatively cool air is necessary since the material from which the turbine blades are made would not be able to withstand the temperature of the undiluted hot products of combustion, or the corrosive effects of the hot ash from residual fuels. The gases pass through the turbine and rapid expansion causes rotation of the shaft. The turbine supplies mechanical power to drive the compressor and auxiliaries, and the process continues as long as fuel is burned. The compressor and auxiliaries absorb about 60% of the total energy supplied as fuel to the combustion chamber.

The gas turbines used for aircraft propulsion are of two types:
  (i) The true jet engine (turbo-jet) (Fig. 29), in which the turbine drives the compressor and auxiliaries, and the hot gases leaving the turbine discharge through the tail nozzle at a velocity in the region of 1700 ft/sec and propel the aircraft in accordance with the principle that every action has an equal and opposite reaction.
 (ii) The turbo-prop (Fig. 30), in which the energy of the hot gases is absorbed by a turbine to drive the compressor and also a propeller.

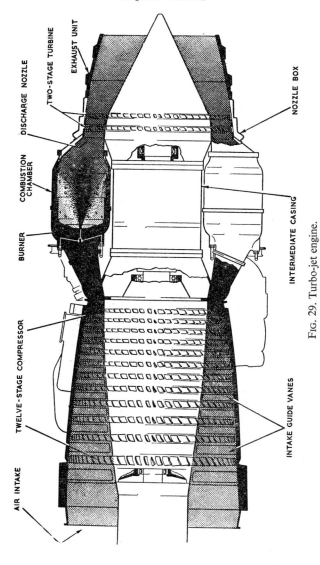

Fig. 29. Turbo-jet engine.

LIQUID FUELS IN BOILERS, FURNACES AND TURBINES 111

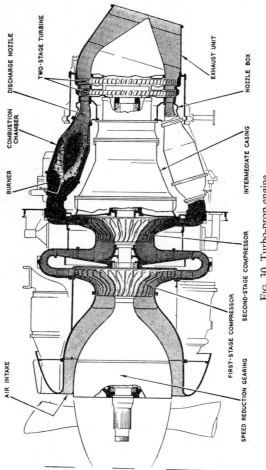

Fig. 30. Turbo-prop engine.

Turbo-jet engines operate best at high altitudes, since the reduction of air temperature improves the efficiency of the compressor, and the reduced temperature and pressure of the atmosphere produce increased turbine efficiency. It is therefore more economical to operate turbo-jet engines at high altitudes, say above 30,000 feet. The efficiency of a propeller or airscrew decreases at higher altitudes and in general turbo-prop engines operate at altitudes of 20,000 to 30,000 feet, and are consequently used for short and medium distances.

Since the thermal efficiency of a gas turbine is lower than that of a steam turbine or diesel engine, the fuel economy is poor. Nevertheless the ease of installation and low maintenance and lubrication costs of the gas turbine make them attractive where fuel economy is not of prime importance. They have been applied to locomotive and marine propulsion, sometimes in the latter application in conjunction with diesel engines which are used for manoeuvring, the gas turbine being used for full power output requirements. Gas turbines have been installed in power stations for meeting peak loads,[19] their low capital cost offsetting their higher fuel consumption compared with steam turbine sets. Other applications have been in oilfields and oil refineries, where fuel economy might be of secondary importance.

Liquid Fuels for Gas Turbines

In considering liquid fuels for gas turbines it is convenient to divide the applications into aviation machines and industrial machines, since whereas the latter require the cheapest grade of fuel, aircraft gas turbines require more stringent specifications.

Fuels for aviation gas turbines.—The early gas turbines used in aircraft used kerosine as the fuel, but since the quantity of kerosine available in any crude oil is limited and an increased availability of the fuel became necessary, it was essential to widen the fraction used as aviation fuel. Thus aviation turbine gasoline (wide cut gasoline) was introduced, this being a mixture of kerosine with some of the components of motor gasoline. The main characteristics

LIQUID FUELS IN BOILERS, FURNACES AND TURBINES 113

required of an aero gas turbine fuel are as follows:[20]

(a) The fuel must flow freely to the engine under all operating conditions. Thus the freezing points must be as low as −60°C, and limits are imposed on vapour pressure to avoid vapour locking in the fuel system. The water content may give rise to clogging or icing of filters or other parts of the fuel system. At the present time it is considered impracticable to remove dissolved water which, provided it remains dissolved, has no adverse effect upon engine operation. It is desirable to limit the undissolved water content of aviation turbine fuels to 30 ppm maximum, and an instrument called the BP Aquascan has recently been developed to determine such small quantities on a continuous basis while the aircraft is being fuelled. Measurement is based on a light absorption technique and the fueller is automatically shut down if the undissolved water contamination exceeds a specified figure.

(b) Combustion must be efficient and stable under all conditions. The fuel may be sprayed into the combustion chamber in an atomized form or may be preheated to enter as a vapour. The stability and efficiency of combustion will depend largely on such factors as combustion chamber temperature and design, turbulence and recirculation patterns, quality of atomization, and the evaporation rate of the fuel, the latter being related to volatility and latent heat. The physical properties of the fuel, such as viscosity and surface tension, affect the quality of atomization. Carbon formation is undesirable, since this may result in erosion of the turbine blades, blockage of the turbine nozzles, or deposition in flame tubes, with consequent loss of combustion efficiency and possible buckling of the flame tubes. Aromatics burn with a smoky flame containing incandescent carbon particles, so that the aromatic content of aviation turbine fuels is limited in the specifications. The end point of the fuel is specified since an increase in the end point would tend to cause increased carbon formation.

(c) The calorific value should be as high as possible. Whether this property is expressed on a weight basis or on a volume basis depends on the type of operation. Where the volume carrying capacity is limited, as on a fighter aircraft, the maximum calorific value on a volume basis is required, that is, a fuel of high specific gravity. Where the payload is important, calorific value on a weight basis is the criterion, that is, a fuel of low specific gravity.

The aero gas turbine can perform satisfactorily on a wide range of distillate fuels, provided good combustion equipment is used. Typical inspection data for two common aviation gas turbine fuels are given in Table 26.

It will be seen that, whereas one of the fuels is of the kerosine type, the other is the so-called wide-cut gasoline, which is manufactured by blending a gasoline fraction with kerosine. Civil aircraft now use fuels of either type, and the wide-cut gasoline is the main fuel for military aircraft. Considerable controversy exists in regard to the relative fire hazards associated with the two types of fuel.

Fires can be caused by ignition of (a) a flammable mixture of fuel vapour and air in the fuel tank, (b) foam generated in the fuel tank during flight, (c) bulk fuel or a spray of fuel resulting from a crash, or (d) flammable vapours in the vicinity of an aircraft during fuelling. A recent report of a Working Party of the Ministry of Aviation[21] has assessed these potential hazards, and the findings are summarized in Table 27.

The source of ignition in fuel tanks could be lightning or static electricity, but only one known case is recorded of such ignition during flight. Fires have occurred during fuelling and are known to have been due to static electricity, but in these cases non-standard equipment was used. There is no known instance of a fire occurring during fuelling where all the correct fuelling regulations have been followed.

TABLE 26

Typical Inspection Data for Aviation Gas Turbine Fuels

	Aviation turbine gasoline	Aviation turbine kerosine
Specific gravity, 60/60°F	0·775	0·800
Distillation, °C, 10%	125	180
20%	135	196
50%	180	228
90%	235	258
FBP	280	285
Flash point, °F	—	130
Freezing point, °C	−60°C	−50°C
Aromatics, % vol	10	17
Appearance	Visually clear, bright, and free from sediment, suspended matter, and undissolved water at normal ambient temperature	

TABLE 27

Summary of the Relative Risks with Aviation Kerosine and Aviation Turbine Gasoline

Fire hazard	The safer fuel
Flammable mixture of fuel vapour and air in fuel tank	Kerosine
Flammable foam generated in fuel tank during flight	Gasoline
Discharge of bulk fuel resulting from crash	Kerosine
Discharge of spray of fuel resulting from crash	Both fuels similar
Discharge of flammable vapours during fuelling	Kerosine

Fuels for industrial gas turbines.—For industrial gas turbines economical running can best be achieved by the use of the cheapest, (i.e. residual) fuel. Distillate fuels present few problems in gas turbines, but residual fuels require more favourable conditions for combustion and lead to difficulties resulting from the constituents of the ash.

Preheating is used to reduce the viscosity of the fuel to a level at which it can be atomized satisfactorily with blast atomizers or pressure jets, and the combustion chamber should be at a fairly high temperature in order to achieve rapid combustion. Refractory-lined or all-metal air-cooled flame tubes are used. It is also desirable to use appreciable quantities of excess air when burning residual fuels since this promotes rapid combustion and helps to reduce the heat radiation to the flame tube.[22] Frequently heat exchangers are used to preheat the combustion air by the hot products of combustion.

The small quantity of ash which is present in residual fuels (usually below 0·05% wt. for turbine fuels) forms deposits on the turbine blades and, at a temperature in excess of 1200°F, has a corrosive effect on the blade metal, both effects resulting in loss of power. The constituents responsible for these troubles are the compounds of sodium and vanadium, and although sodium can be reduced by centrifuging or washing the fuel, no commercial method is available for removing vanadium. The effect of various additives, such as kaolin, Epsom salts, and metallic additives such as magnesium, zinc and barium, in reducing deposit formation and corrosion, have been studied, but although the results are promising, none of these additives has so far proved ideal in suppressing corrosion and deposition. The most effective commercial additive to date is ethyl silicate. One of the major difficulties encountered with additives is their cost, since unless an additive is cheap the economic advantage of using a residual fuel is lost. Controlled combustion to produce carbon in the combustion products has been shown[23] to alleviate deposit formation and corrosion, the mechanism being either that the carbon particles encase the soft, sticky ash particles, or that the comparatively low melting V_2O_5 is reduced by carbon to the higher melting V_2O_3.[24] This production of carbon results in a combustion loss which does not exceed 1%. It is desirable that the ash be 50% soluble in water in order that the turbine blades can be easily washed.

References

1. *U.K. Petroleum Industry Statistics*, Petroleum Information Bureau (1961)
2. W. B. CARLSON, *Proc. Symposium on "Flames and Industry"*, Int. Fuel, p.B-10 (1957)
3. K. E. DADSWELL and F. R. THOMPSON, *Proc. Conf. on Major Developments in Liquid Fuel Firing*, Inst. Fuel, p.D-1 (1959)
4. K. J. MACKENZIE et al., *The Engineer*, **210**, 548 (1960)
5. C. A. ROAST, *Proc. Conf. on Major Developments in Liquid Fuel Firing*, Inst. Fuel, p. B-1 (1959)
6. G. WHITTINGHAM, *Proc. Conf. on Major Developments in Liquid Fuel Firing*, Inst. Fuel, p. B-19, (1959)
7. E. BRETT DAVIES and A. B. PRITCHARD, *Proc. Conf. on Major Developments in Liquid Fuel Firing*, Inst. Fuel, p. B-37. (1959)
8. M. RODDAN, *Proc. Conf. on Modern Applications of Liquid Fuels*, Inst. Fuel (1948)
9. J. A. GRANGER, *J. Iron and Steel Institute*, **194**, 307 (1960)
10. M. RIVIERE, *J. Inst. Fuel*, **30**, 556 (1957)
11. F. A. GRAY et al., *Proc. Conf. on Major Developments in Liquid Fuel Firing*, Inst. Fuel p. A-18 (1959)
12. A. HOGG and C. HOLDEN, *Proc. Conf. on Major Developments in Liquid Fuel Firing*, Inst. Fuel, p. A-53. (1959)
13. C. HOLDEN and M. W. THRING, *Proc. Conf. on Major Developments in Liquid Fuel Firing*, Inst. Fuel, p. A-60 (1959)
14. M. W. THRING, *J. Iron and Steel Institute*, **173**, 381 (1956)
15. W. R. BULCRAIG, *Proc. Conf. on Major Developments in Liquid Fuel Firing*, Inst. Fuel, p. B-52 (1959)
16. S. KRUSZEWSKI, *Proc. Symposium on "Flames and Industry"*, Inst. Fuel, p. B-5 (1957)
17. K. C. SHARP, *J. Iron and Steel Institute*, **199**, 69 (1961)
18. P. HAZARD, *J. Iron and Steel Institute*, **199**, 127 (1961)
19. Anon., *Boiler House Review*, **78**, (3), 78 (1963)
20. C. G. WILLIAMS, *J. Inst. Petroleum*, **33**, 267 (1947)
21. Report of the Working Party on Aviation Kerosine and Wide-Cut Gasoline, HMSO (1962)
22. R. F. DARLING, *J. Inst. Fuel*, **32**, 475 (1959)
23. A. T. BOWDEN et al., *Proc. Inst. Mech. Engrs.*, **167**, 291 (1953)
24. P. T. SULZER, *Schweizer Archiv.*, **20**, 33 (1954)

CHAPTER 8

The Use of Liquid Fuels for Gas Manufacture

THE USE of petroleum oils in the gas industry is not new, since gas oil was used for enriching blue water gas before the end of the nineteenth century. During the past decade, however, the use of oil for gas manufacture has expanded rapidly from a peak load usage to one which can economically take a large proportion of the base load. Oil gasification processes are now competitive with the more conventional solid fuel processes and are advantageous in regard to feedstock handling, plant size, and the elimination of ash and clinker. Furthermore, the economics of conventional gas making plant based on the carbonization of coal requires a good market for coke. The use of oil as feedstock reduces the coke-to-gas balance, a feature which the gas industry welcomes at a time when there are indications of a market resistance to solid fuel. The Lurgi complete gasification process for the gasification of coal is attractive for the same reason.

The developments in oil gasification processes resulted from the difficulties encountered in the coal industry after the second world war. Coal supplies became difficult, the proportion of lower grade coals increased, and the price of coal increased at a faster rate than that of oil.

A number of established processes are now available to produce, from a wide range of petroleum feedstock, suitable gases for distribution as town gas, for the replacement of producer gas for industrial use, or for synthesis gas for the chemical industry. In general, the fuels used for the manufacture of gas are light distillate, gas oil, medium and heavy fuel oils, and refinery tail gases. Typical properties of these feedstocks are given in Table 28. Liquefied petroleum gases, propane and butane, are too expensive for use in the town gas industry except in isolated cases, and their

TABLE 28

Typical Properties of Feedstocks Used in Oil Gasification Plants

	Light distillate	Gas oil	Medium fuel oil	Heavy fuel oil
Specific gravity 60/60°F	0·70	0·84	0·95	0·965
Flash point, °F	Open	150	210	230
Sulphur, % wt	<0·1	0·8	2·3–3·5	Up to 4·5
Viscosity at 100°F, Redwood 1 sec		35	500–900	3000
Conradson carbon residue, % wt		0·05 (10% bottoms)	4–7	10–12
Gross calorific value B.T.U./lb	20,400	19,600	18,400	18,300

	Refinery gas		
	Lower range	Medium range	Upper range
Specific gravity (air = 1)	0·77–0·91	0·92–1·23	1·19–1·39
Gross cal. value, B.T.U./ft^3	1300–1500	1500–2000	2000–2240
Composition, % vol			
Nitrogen	0·1–4·1	0·0–11·7	0·0–10·1
Hydrogen	28·2–39·1	6·1–31·8	2·9–15·3
Carbon monoxide	0·0–0·8	0·0–1·2	0·0
Carbon dioxide	0·0–1·3	0·0–2·0	0·0–1·0
Hydrogen sulphide	0·1–2·3	0·0–3·2	0·1–1·0
Methane	6·8–17·5	11·1–22·9	12·9–16·2
Ethylene	2·0–3·4	trace–4·9	1·8–5·7
Ethane	10·8–14·1	10·3–22·1	11·0–19·0
Propene	6·7–11·5	2·7–14·9	5·0–15·8
Propane	14·8–20·6	12·1–28·4	17·7–25·5
Butane	4·3–9·3	1·1–23·1	18·3–30·4
Pentane		0·0–2·6	0·0–2·1
Butene		0·0–3·7	0·0–9·9

use is confined to the domestic and commercial field and for particular purposes in industry.

Characteristics of Town Gas and Industrial Gases

Town Gas

Town gas is distributed in pipes to industrial and domestic consumers, and the standard of quality is controlled by law, each gas undertaking producing to a declared calorific value. This property alone does not provide a complete safeguard against variation in combustion characteristics and hence against unsatisfactory burner operation. For this purpose a further property is specified, namely the Wobbe Index, which is defined as the ratio of the calorific value to the square root of its specific gravity. The characteristics of town gas are given in Table 29.

TABLE 29

Typical Characteristics of Town Gas

Component	% volume
CO_2	1·5–3·5
CO	7·5–15·5
H_2	49·0–57·0
C_nH_{2n+2}	21·0–30·0
C_nH_m	1·5–3·5
O_2	0·2–0·6
N_2	3·0–9·0
B.T.U./ft^3	470–560
Sp. gr. (air = 1)	0·38–0·48
Wobbe Index	700–850

Producer Gas

Producer gas is manufactured by the action of air and steam on coal or coke, and finds its main application in the production of iron and steel, glass, refractories, ceramics, and in the carboniza-

THE USE OF LIQUID FUELS FOR GAS MANUFACTURE 121

TABLE 30

Typical Characteristics of Cold Clean Producer Gas

Component	% volume
CO	25–30
H_2	10–18
N_2	50–55
CH_4	0·5–2·5
C_nH_m	0–0·5
CO_2	3·0–6·0
O_2	0·2–0·4
B.T.U./ft^3	135–165
Sp. gr. (air = 1)	0·87–0·90

tion industry. The gas consists essentially of carbon monoxide, hydrogen, and nitrogen, enriched when the fuel is coal by the products of distillation.

Synthesis Gas

Synthesis gas is used for the manufacture of ammonia and such chemicals as methanol. The use for which the synthesis gas is required determines its composition; ammonia synthesis requires a mixture of three parts of hydrogen to one part of nitrogen, and other syntheses generally a mixture of two parts of hydrogen to one part of carbon monoxide.

Principles of Oil Gasification

The feedstocks used in oil gasification all have C/H ratios which are higher than those of the gases which it is required to produce. Town gas has a C/H ratio of approximately 3 : 1, whereas a heavy petroleum oil has a ratio of about 7·8 : 1. The conversion of liquid to gas can therefore only be carried out by the removal of carbon or by the reaction of the hydrocarbons with steam and/or hydrogen.

The processes used in the gasification of petroleum products are:
(a) Thermal cracking with steam, with and without catalysts.
(b) Partial combustion.
(c) Hydrogenation.

Thermal Cracking

When hydrocarbon oils are heated, cracking occurs. That is, large vapour molecules are thermally decomposed to smaller ones of lower boiling points and finally to gas. The reactions involved vary with the type and size of the hydrocarbon molecules, with the conditions of temperature and pressure, and also with the presence or absence of catalysts.

Thermal cracking causes lower paraffins to form hydrogen and deposit carbon, whereas the higher paraffins tend to crack into a paraffin and an olefin. Polymerization of olefins and complex reactions with naphthenes and aromatics result in the formation of tar. In general, moderate temperatures and high pressures cause cracking of carbon-to-carbon bonds near the centre of the chain, whereas high temperatures and moderate pressures tend to break the chain nearer the ends and cause rupture of carbon-to-hydrogen bonds. Thus the latter conditions favour the production of larger proportions of permanent gases and carbon.

Catalysts can be used to accelerate the reaction rates and also to influence the reaction so as to produce more gaseous products.

When steam is used in conjunction with cracking, reactions occur between the steam and the hydrocarbons to produce carbon monoxide and hydrogen. With no catalyst, high temperatures are necessary and lead to the formation of carbon black and a smaller amount of tar. When the reactions are catalysed, lower temperatures can be used and a certain amount of tar is formed.

Partial Combustion

Oxygen, if in sufficient quantity, reacts with hydrocarbons under suitable conditions of temperature to give the products of complete combustion, namely carbon dioxide and water vapour. With less than the stoichiometric quantity of oxygen the reactions are more complicated and the products depend on the temperature, the type of hydrocarbon, and the degree of chemical equilibrium attained. The reactions are exothermic and usually cracking occurs simultaneously with partial oxidation, the processes thus being "autothermic" and continuous in operation. Since reactions with steam are endothermic, steam can be used to control the temperature in partial combustion processes. The steam, together with the oxygen, forms oxides of carbon and so reduces the amount of carbon or tar which would otherwise have been produced from a pure cracking reaction.

The gas produced from partial combustion with air has a high specific gravity owing to its high nitrogen content, and this severely limits its use as an additive to town gas. When oxygen is used the resultant gas does not suffer this disadvantage and can be used as an additive to town gas to meet peak loads. These processes are used chiefly for the production of synthesis gas and also for hydrogen-rich gas for hydrogenation processes.

Hydrogenation

In the two methods of gasification given above the C/H ratio of the feedstock is reduced to that of the finished gas either by the removal of carbon or by the action of steam. Hydrogenation processes reduce the C/H ratio by the direct action of hydrogen at temperatures of 1300–1650°F and pressures of approximately 30 atm, the reactions being exothermic.

Town Gas Manufacture

Carburetted Water Gas

The original use of petroleum products in the gas industry was for the enrichment of blue water gas by the products obtained from the thermal cracking of a middle distillate oil (boiling range 200–300°C). In this way a gas suitable to supplement the usual coal-gas type of town gas was produced.

The plant consists of a blue water gas generator, a carburettor, a superheater, and a waste heat boiler (Fig. 31). The carburettor is connected to the gas offtake from the generator, and provision is made for the introduction of secondary air and of enriching oil. At the top of the superheater is the gas offtake to the wash box and a connection to the waste heat boiler.

The hot exhaust gas from the air blow is a lean producer gas containing approximately 10% CO, which is burned with secondary air in the carburettor. The products of combustion leave the

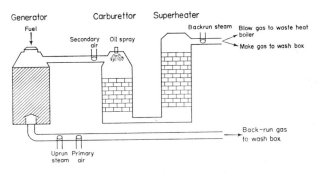

FIG. 31. Carburetted water gas plant.

carburettor and raise the temperature of the superheater before passing to the waste heat boiler.

The gas making run is then commenced, oil being sprayed into the top of the carburettor immediately after the opening of the steam valve. The oil is vaporized and cracked in the carburettor, the mixture of blue water gas and thermally cracked gas oil passing through the superheater to complete the cracking operation. Carbon deposited in the carburettor and superheater is burned during the air blow. The up-run of steam is followed by a back-run, in which steam is admitted from the top of the superheater through the carburettor and fuel bed, the back-run gas being led through a connection from the bottom of the generator to the wash box. The back-run is followed by a short up-run steam purge which precedes the blow.

The back-run reduces the sensible heat loss in the make gas, since the gas and undecomposed steam leaving the bottom of the generator are at a temperature of about 800°F instead of the 1200–1400°F at which the up-run gas leaves the superheater. The back-run steam also cools the top of the superheater to enable more heat to be extracted from the up-run gas and blow products.

Oil injected during the back-run is vaporized and carried with the steam through the fuel bed, where it is subjected to severe cracking (reforming), as distinct from the more moderate cracking which takes place in the carburettor and superheater during the up-run. Fuel bed reforming can also be carried out during the normal up-run. In this sense the term "reforming" must be distinguished from the more accepted use of the word in petroleum technology (see Chapter 1).

The oil gas obtained from pure reforming has a calorific value of about 500 B.T.U./ft^3 and a specific gravity of approximately 0·40, whereas the gas obtained from the less severe cracking in the carburettor has a calorific value of 1500 B.T.U./ft^3 and a gravity of approximately 0·95. This high B.T.U. gas, when mixed with blue water gas, produces 500 B.T.U./ft^3 gas with a gravity of 0·5–0·55.

Feedstocks ranging from light distillate to heavy fuel oils may be used in carburetted water gas sets. The cost differential has encouraged research into methods of using residual fuel oils or light distillate fuels in place of the conventionally used gas oil. In addition to heavier deposits of carbon in the carburettor, residual fuel oils of high sulphur content (say 4%) lead to the formation of carbon disulphide, which is not removed in ordinary methods of gas purification. Back-running steam reacts with the carbon deposits, keeps the surfaces clean, and reduces the amount of carbon disulphide formed. When heavy oil is used for enriching, about 50% of the oil is sprayed into the top of the generator, with the result that residual carbon deposited on the fuel bed is gasified along with the coke. The remainder of the oil is added to the carburettor. Preheating to approximately 240°F and pressures of 150–250 psig are necessary when heavy fuel oils are used, and checkerless carburettors are preferred in order to facilitate cleaning. The yields obtained in modern plants are about 1·3 therms per gallon of gas oil and 1·1 therms per gallon of heavy fuel oil.

Non-catalytic Steam Cracking Processes

The new Jones oil gasification process.—In the Jones process, which is cyclic, oil is thermally cracked with steam at temperatures

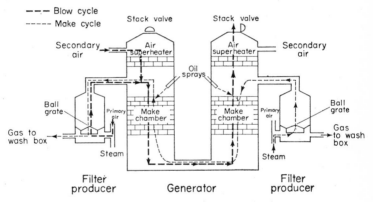

Fig. 32. New Jones oil gasification process.

sufficiently high to produce a gas having a calorific value similar to or lower than town gas. Appreciable quantities of carbon black are produced, together with smaller quantities of tar, and from an economic viewpoint, it is usually necessary to have an attractive market for carbon black.

The plant (Fig. 32) consists of a generator and two filter producers. The generator is divided into two vessels which are connected at the bottom. Each vessel has a stack valve and contains an air superheater section and a make chamber, both containing checkerwork. Oil sprays are situated below the gas offtake and above the main checkerwork in the generator. The combined gas offtake and blast connection lead to a filter producer, at the base of which is the gas outlet connection to the wash box, together with primary air and steam inlets. The principle is similar to that of the carburetted water gas process, except that no coke is used, the producer gas and water gas being obtained from the carbon formed from cracking the feedstock. Since carbon formation is essential to the process, the feedstock must have a Conradson carbon residue of more than 6% wt.

Each cycle consists of a blow and run in each direction. Primary air is blown through, say the left hand filter producer, the resultant producer gas being burned with preheated secondary air above and through the make checkerwork. During this stage the right and left hand gas outlet valves are closed, as is the left hand stack valve; the right hand stack valve is open. The hot products pass through the right hand make and preheat checkerwork and out to atmosphere via the stack. The left hand producer and both generator vessels have now been heated to temperature, and the run or make stage follows. Steam is introduced to the left hand producer, the resulting blue water gas and undecomposed superheated steam passing to the left hand make checkerwork, on which oil is sprayed. The oil vapours crack in the atmosphere of blue water gas and steam whilst passing through the make checkerwork in both right and left hand vessels, then pass through the filter producer to the wash box. Roughly half the suspended carbon black is filtered out in a bed of mechanically agitated refractory spheres and is then consumed when the blow and run cycle is repeated from right to left. The remainder of the carbon, equivalent to 20% of the fuel used, is removed in the wash box and can be used for steam raising. In modern plants one blow and run cycle normally takes 7–10 min.

The gas produced has a calorific value which depends on the degree of cracking; 360 B.T.U./ft^3 with severe cracking and maximum quantities of carbon black to 520 B.T.U./ft^3 under less severe conditions (Table 31). The latter gas is completely inter-

TABLE 31

Characteristics of Gases Produced in the New Jones Oil Gasification Process using Heavy Fuel Oil of Conradson Carbon 11% wt.

Component	% volume	
H_2	49·8	62·5
C_nH_{2n+2}	22·8	12·9
C_nH_m	3·3	0·5
CO	15·5	7·4
CO_2	3·2	4·6
O_2	0·6	0·2
N_2	4·8	11·9
B.T.U./ft^3	520	360
Sp. gr. (air = 1)	0·46	0·38
Gas make, ft^3/ton	51,750	43,000
Efficiency, therms gas/therm oil	65%	38%

changeable with normal town gas, whereas the former is convenient as a diluent for denser carburetted water gas or high B.T.U. gas produced in the Hall or similar process.

The Hall "high B.T.U." oil gasification process.—The Hall type of process is cyclic and non-catalytic, and was evolved to produce from high Conradson carbon fuel oils a rich gas to supplement or substitute for natural gas. The degree of cracking is much less severe than in the Jones process, temperatures of 1450–1650°F being used to produce a range of calorific value, generally 900–1300 B.T.U./ft^3. Steam is used but plays little part in the reactions, acting mainly as a vehicle for hydrocarbon vapours. The necessary change in C/H ratio is brought about almost entirely by the deposition of carbon and the production of tar. Thus the gas yields using medium fuel oils do not exceed 60% of the thermal value of the feedstock, and the overall efficiency (gas plus tar) is approximately 80%.

Distillates and residual oils can be used as feedstock, although gasification efficiency and output fall away considerably when the Conradson carbon exceeds 8% wt. To avoid excessive carbon and tar formation with high Conradson carbon feedstock it is normal to make a gas of 1100–1300 B.T.U./ft^3, which is then diluted by passing the first blow gases into the make (blow running) to reduce the calorific value to 900–1000 B.T.U./ft^3. To attempt to produce a gas in excess of 1300 B.T.U./ft^3 would result in the

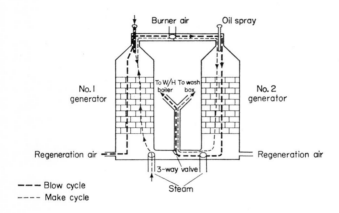

Fig. 33. Hall "High B.T.U." process.

separation from the gas of uncracked oil. The gas produced is high in olefin content, which is usual with oil gases of high calorific value.

The application of these high B.T.U. gases is limited in the UK, the main advantages being low capital costs (some redundant carburetted water gas plants have been suitably modified) and the production of a gas suitable to be mixed with, say low calorific value Jones gas or hydrogen from a Shell type process for town gas distribution.

The plant (Fig. 33) consists essentially of two generators containing checker brickwork. The generators are connected at the top and have connections at the base for regeneration air and steam. The base of each generator is connected by means of a 3-way valve to the stack via a waste heat boiler, and to the make gas pipe and wash box. At the top of each generator is a burner air connection and a combined make-oil/heating-oil spray.

A cycle is as follows. The regeneration air is admitted to the base of No. 1 generator, burns off the carbon deposited during the previous make and restores the temperature throughout the system. The oil burner can be used to augment the heat production. The waste gas passes from the base of No. 2 generator via the waste heat boiler to the stack. The air supplies are closed and a steam purge enters the base of No. 1 generator. The 3-way valve is reversed and, with the steam still flowing, No. 1 make oil opens. The steam is preheated by the checkers and meets the oil spray, the oil and steam passing downwards through No. 2 checkerwork to complete the thermal cracking. During this stage the gases pass via the make gas pipe to the wash box. The final stage is that No. 1 make oil is shut off and the steam flow continued to purge the plant.

TABLE 32

Characteristics of Typical Gas Produced in the Hall Process

Component	% volume
H_2	19·4
CH_4	28·1
C_2H_6	6·6
C_nH_m	23·4
CO	1·6
CO_2	4·7
O_2	0·9
N_2	15·3
B.T.U./ft^3	1050
Sp. gr. (air = 1)	0·83

The first four stages are then repeated in the reverse direction. A typical blow and make in one direction takes approximately four minutes, the total cycle being of eight minutes duration.

The Semet Solvay process and the Gasmaco process are both of the Hall type, but differ somewhat in the plant design and in the manner in which the oil is vaporized and cracked.

Catalytic Steam Cracking Processes

The thermal cracking processes already described produce carbon in large quantities, with consequent low gas yields and efficiency. The Jones process produces excessive carbon and its economy depends on a use for carbon black. Whereas the Hall type processes do not result in the formation of surplus carbon, the resultant gas has a high content of unsaturated hydrocarbons and is unsuitable for blending to any great extent in town gas.

In catalytic processes a suitable catalyst is used to promote the steam/carbon reactions and possibly reactions between steam and gaseous hydrocarbons, with the result that tar and carbon formation are reduced and gas yields are considerably increased. In purely thermal cracking processes steam reacts to a smaller extent, as shown by the lower oxygen contents of the product gases. Catalytic processes are the most suitable for the town gas industry at present owing to their high efficiencies and flexibility in regard to the product gas and feedstock.

The South Eastern Gas Board developed the Segas process, which uses a lime catalyst to facilitate the reactions. The work of two French companies resulted in the Onia-Gegi process, which is based on a sulphur-resistant nickel based catalyst. These two cyclic catalytic processes, which manufacture a gas similar to town gas, are essentially similar, differing mainly in the type of catalyst employed.

In both the Segas and Onia-Gegi processes the sulphur content of the gas is considerably less than that of crude coal gas and of the gas obtained from the same oil by thermal cracking, since a large proportion of the sulphur in the oil is fixed by the catalyst during the make and released to the waste gases during the blow.

As distinct from these two cyclic processes, the Hercules catalytic process, which operates on gaseous or light liquid feedstocks in an externally heated bed of catalyst, is continuous.

The Segas process.—Feedstock varying from liquid butane to residual fuels of 10% Conradson carbon can be used, and the calorific value of the gas is determined principally by the temperature of the catalyst and to a lesser extent by the type of feedstock employed. Some tar is produced when gas oil or fuel oil is gasified, but when light distillate is used only fine carbon is collected in the washer. The CO_2 content of the gas increases with the specific

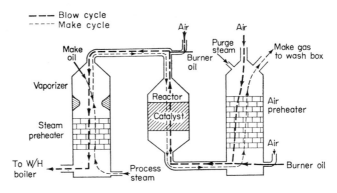

FIG. 34. Segas three-vessel process.

gravity of the feedstock, and thus when using heavy fuel oil the gas is not completely interchangeable with normal town gas.

Figure 34 shows a three-vessel Segas plant. Two-vessel plants have been built in which the catalyst chamber and air preheater are contained in the same vessel.

The method of operation of the three-vessel type of plant is that counterflowing air is blown through the air preheater and burns off any carbon. The preheated air brings the catalyst to the working temperature of 1450–1800°F at the beginning of the run,

THE USE OF LIQUID FUELS FOR GAS MANUFACTURE 133

and heats up the vaporizer and steam preheater, burning off any carbon in its path. Finally the waste gases are exhausted through the waste heat boiler to atmosphere. A steam purge follows in the same direction. During the make stage the stack valve is closed and the gas valve to the wash box opened. Steam is admitted in the opposite direction to the blow and is preheated before passing to the vaporizing chamber to meet the counter-current oil spray. The restriction above the steam preheater creates a high velocity and prevents oil droplets from falling on to the checkerwork below. The mixture of steam and vaporized oil passes through the catalyst bed, where the desired reactions occur. Sensible heat in the gas is recovered in the air preheater through which it passes to the wash box. A purge follows the make, and the cycle is repeated.

TABLE 33

Typical Plant Data for the Segas Process

	Feedstock		
	Heavy fuel	Gas oil	Light distillate
Process oil, gal/1000 ft^3 gas	3·91	3·60	4·32
Heating oil, gal/1000 ft^3 gas	0·18	0·20	0·40
Steam, lb/1000 ft^3 gas	55·0	47·7	35·3
Air, ft^3/1000 ft^3 gas	3250	3170	2320
Tar produced gal/1000 ft^3 gas	0·46	0·20	0
Gas analysis, % volume			
CO_2	9·2	5·8	3·0
O_2	0·5	0·4	0·8
C_nH_m	6·6	4·2	5·1
CO	14·8	8·3	15·1
H_2	49·1	50·3	57·6
CH_4	15·6	16·2	14·7
N_2	4·2	4·8	3·7
Calorific value, B.T.U./ft^3	500	492	490
Sp. gr. (air = 1)	0·56	0·52	0·40

Carbon deposition during the make is not sufficient to restore the temperatures throughout the system to the desired levels during the blow. Oil tar is normally used to supplement the carbon, the amount varying from approximately 5% with heavy oil to 10% when using light distillate feedstock. The plant is automatically controlled and the cycle varies from three to six minutes.

The Onia-Gegi process.—The three-vessel Onia-Gegi process is similar to the Segas process. There is also available a two-vessel non-regenerative process in which the blow and make stages proceed in the same direction.

Industrial Gas Manufacture

Partial Combustion Processes

Gases made by partial combustion processes using air have high nitrogen contents and hence high specific gravities. They are thus generally unsuitable for distribution with normal town gas. Such processes include the Dayton process, the similar Gas, Light and Coke Co. process, the Koppers-Hasche process, and the Distrigas process, all of which suffer from the added disadvantage of producing little reduction in the C/H ratio of the feedstock.

In the more recent Shell and Texaco processes gaseous or liquid hydrocarbons are gasified with oxygen and steam under pressure. These processes were developed for synthesis gas manufacture, but recently the Shell process has been incorporated in a town gas installation, where the water gas produced is processed for removal of carbon monoxide, carbon dioxide and sulphur, and the residual hydrogen is blended with either high B.T.U. gas from Segas plant or with refinery tail gas, and ballasted with nitrogen to give a town gas of the required characteristics.

Partial combustion processes are most suitable for the direct firing of industrial plant. Gaseous combustion is sometimes preferred to the direct firing of oil, for example in cases where a soft, lazy flame is required for a metallurgical furnace, or where furnace atmosphere control is of prime importance. Below are given descriptions of some of the existing types of plant.

The Dayton process.—This process, in which a vaporized distillate is partially combusted with air at 1300–1650°F, is continuous and non-catalytic. The generator is shown in Fig. 35, and data for the gasification of a light fuel oil are given in Table 34.

Preheated air and atomized oil are fed into the top of a heavy steel tube, which is externally heated by the reacted gases and projects almost to the bottom of a refractory lined cylindrical

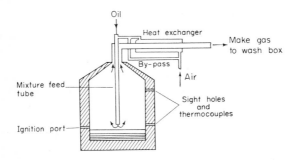

Fig. 35. Improved Dayton process.

chamber. The vapours ignite at the base of the generator and the product gases pass through a heat exchanger, which acts as the air preheater, before being water washed to remove tar and oil. The air used is about 10% of that required for complete combustion and the preferred feedstocks are distillates or light residual fuel oils.

The throughput of the plant is high, the capital cost is low, and full gas making capacity can be achieved in less than one hour.

The Dayton–Faber process.—This is a development of the Dayton process, oxygen-enriched air being used in a modified form of generator to produce gases containing less nitrogen and of calorific value 500–900 B.T.U./ft^3.

TABLE 34

Typical Characteristics of Gas Produced in the Dayton Process

Generator temperature, °F	1350	1500	1650
Gas analysis, % volume			
CO_2	6·1	5·4	3·6
O_2	0·5	0·5	0·3
C_nH_m	15·9	12·0	6·6
CO	5·5	8·7	11·4
H_2	1·5	3·9	12·0
C_nH_{2n+2}	8·7	7·6	10·8
N_2	61·8	61·9	55·1
Calorific value, B.T.U./ft^3, cold, clean gas	500	400	300
Sp. gr. (air = 1)	1·03	1·00	0·85
Gas yield, therms gas/therm oil	0·79	0·80	0·82

The Shell gasification process.—This consists of the controlled, non-catalytic, partial oxidation of any hydrocarbon feedstock by the use of oxygen and steam under pressures of 10 to 40 atm and at reactor temperatures in the region of 2500°F. A water gas is formed and can be used for synthesis gas, base material for technical hydrogen, or as a component for town gas. Oxygen and feedstock are separately preheated before being mixed with superheated steam and fed to the reactor. A waste heat boiler recovers the bulk of the sensible heat of the gas and carbon is removed in a water wash system, the resultant gas having a carbon content of about 1 part per million. A typical synthesis gas produced from heavy fuel oil would consist of 46% hydrogen, 47% carbon monoxide, 5% carbon dioxide and 0·5% methane.

The Shell process has recently been incorporated in a town gas installation,[9] where it is used for the manufacture of hydrogen, which is then blended with high B.T.U. gas from Segas plant or refinery tail gas, and mixed with nitrogen to give a town gas of

THE USE OF LIQUID FUELS FOR GAS MANUFACTURE 137

the required characteristics. The Shell synthesis gas is processed for sulphur removal and conversion of carbon monoxide to the dioxide, which is subsequently removed under pressure by potassium carbonate lye at about 175°F. The carbon monoxide con-

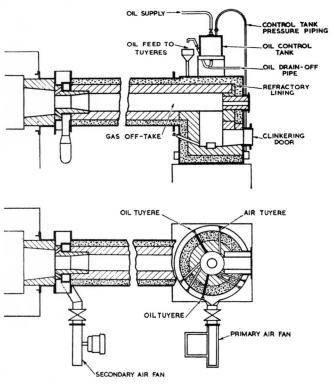

FIG. 36. O.C.C.R. oil gasifier.

version is carried out in catalytic converters, the synthesis gas being saturated with water before high pressure steam is added, after which the mixture is heated to reaction temperature and fed to the reactors. The catalysed water gas shift reaction converts the bulk of the carbon monoxide to dioxide, with the production of hydrogen.

The O.C.C.R. gasifier.—The O.C.C.R. gasifier (Fig. 36) is a partial combustion system wherein the oil is fed on to the base of a refractory lined vertical chamber. Approximately 30% of the stoichiometric air required for complete combustion is supplied as primary air to the gasifier and is distributed around the base to cause partial combustion and cracking to occur. The product gas, together with appreciable quantities of suspended carbon, at a temperature of about 2200°F, passes through a refractory lined duct to the furnace where secondary air is admitted to complete the combustion.

TABLE 35

Gas Analysis and Yields from O.C.C.R. Gasifier

Gas analysis, % volume		
CO_2		4
CO		16
H_2		14
CH_4		4
N_2		62
Calorific value, cold clean gas,	B.T.U./ft^3	120
Sensible heat in gas at 2200°F,	B.T.U./ft^3	50
Potential plus sensible heat in suspended carbon,	B.T.U./ft^3	45
Calorific value, hot raw gas,	B.T.U./ft^3	215
Gross thermal efficiency, hot raw gas, %		96–98

The Urquhart high duty oil gasifier.—The design of this high duty combustion chamber (Fig. 37) is such that the high heat release rates obtained permit fuel oil to be reacted with less than the stoichiometric quantity of air to produce a gas with little or no carbon deposition. It is only with such careful control of combustion conditions and with such high combustion rates as

THE USE OF LIQUID FUELS FOR GAS MANUFACTURE 139

are obtained with this gasifier that partial combustion can be achieved with negligible carbon formation. The gas is of a dry producer type and the reactions can be controlled to provide a range of gases varying in composition and temperature (Fig. 38).

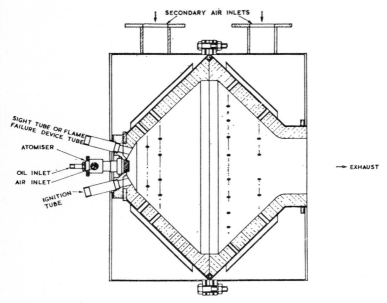

FIG. 37. Urquhart high-duty combustion chamber.

Fuel oil of 3000 seconds can be successfully gasified with three times the fuel/air ratio, the gas containing approximately 5% carbon dioxide, 12% hydrogen, 14% carbon monoxide, and 4% hydrocarbon gases.

Liquefied Petroleum Gases

Liquefied petroleum gases (LPG) are the C_3 and C_4 condensable hydrocarbon gases which have for many years found application for the supply of gaseous fuel to caravans, country houses, and so on, and for the oxy-flame cutting of metals.

These hydrocarbons, gaseous at normal temperatures, can be readily liquefied by the application of moderate pressures and stored or distributed in light pressure vessels. Vaporization occurs by release of the pressure, and the latent heat can be obtained for low flow rates by heat transfer from the surroundings, or by the use of a heated evaporator for high flow rates.

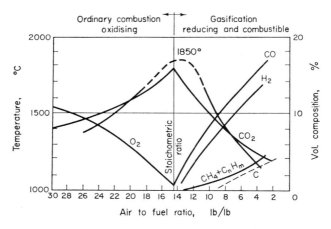

FIG. 38. Composition of combustion products of hot gases from 3500 second Redwood I fuel oil.

LP gases are obtained from wet natural gas and from refinery distillation, cracking and reforming processes. The C_3 and C_4 components are separated by distillation under pressure.

Commercial propane and commercial butane are the two grades of LP gases generally marketed. The quality is controlled to meet specifications for vapour pressure, sulphur content, and dryness. The vapour pressure specification must not be exceeded on account of storage and handling equipment, the sulphur content is 0·02% maximum, and water must be eliminated to

THE USE OF LIQUID FUELS FOR GAS MANUFACTURE 141

TABLE 36

Physical Properties of C_3 and C_4 Hydrocarbons

	Propane C_3H_8	Propene C_3H_6	n-Butane C_4H_{10}	i-Butane C_4H_{10}
Boiling point, °F	−43·8	−53·9	31·1	10·9
Vapour pressure at 70°F, psig	118	147	29	44
Liquid sp. gr., 60/60°F	0·508	0·522	0·584	0·563
Gas sp. gr. (air = 1)	1·546	1·452	2·070	2·066
Ft^3 of gas per lb of liquid, at 60°F	8·62	8·71	6·54	5·27
Gross calorific value, B.T.U./ft^3	2521	2335	3267	3259

avoid freezing of regulators and valves. Since the gases are almost odourless traces of odorizer are generally added.

It will be seen from Table 37 that, compared with town gas, LP gases have high calorific values, high specific gravities, narrow limits of inflammability, and require higher quantities of air for complete combustion. In addition, the flame velocities are lower than those obtained when burning town gas. Owing to these very considerable differences a whole range of burner equipment has been produced specifically for use with LP gases. Approximately 60% of the LP gases produced in the UK are used in "domestic/commercial" equipment, the remainder being used industrially and including gasworks feedstock. The use of LPG is increasing rapidly in the glass, ceramic and metallurgical industries, where its very low sulphur content helps overcome the harmful effects of this element which were previously encountered in these industries.

The Use of LP Gases in Town Gas and Industrial Gas Manufacture

Butane enrichment.—Butane can be used for the enrichment of town gas where it is required to operate continuous vertical retorts with high steaming rates. The coal gas produced under these

TABLE 37

Typical Properties of Commercial Propane and Commercial Butane

	Commercial propane	Commercial butane
Sp. gr. of liquid, 60/60°F	0·509	0·580
Sp. gr. of gas (air = 1)	1·52	2·01
Ft^3 of gas per gallon of liquid, at 60°F	36	31·5
Vapour pressure at 70°F, lb/in^2	124	31
Latent heat of vaporisation at 60°F, B.T.U./lb	152	160
Sulphur content, % wt	0·01	0·01
Limits of inflammability in air,		
Lower limit, %	2·4	1·9
Upper limit, %	9·6	8·6
Calorific value, B.T.U./ft^3	2522	3261
Air required for complete combustion, ft^3/ft^3 of gas	23·4	30·0

conditions has a calorific value of approximately 450 B.T.U./ft^3 and after enrichment is increased to 500–520 B.T.U./ft^3. The addition is limited by the combustion characteristics and high specific gravity of butane.

Butane–air gas.—In isolated areas not connected to a source of town gas, butane–air mixtures of 23% hydrocarbon with air and having a calorific value of about 750 B.T.U./ft^3 and specific gravity of 1·23 can be distributed and used in suitable burners. The composition of the butane–air mixture is well on the rich side of the inflammability limits in order to make the distribution of this gas safe.

LP Gas Installation

An LP gas installation comprises storage vessels, vaporizer, and control equipment. Mixing equipment is provided if an air–gas mixture is to be distributed, and a small gas holder can be used to balance rapid changes in flow rates. The standard sizes of portable storage vessels which can be conveniently transported to a refinery for filling are 1 ton and $2\frac{1}{2}$ tons. The capacity of static storage vessels ranges from 1 to 50 tons, and these are charged from bulk supply road vehicles. For butane plants, evaporators of up to 10,000 lb/hr are employed.

References

1. *Gasmaking*, published by The British Petroleum Co. Ltd. (1960)
2. E. J. LAWTON, The manufacture of town gas from liquid or gaseous fuels, *J. Inst. Fuel*, **30,** 242 (1957)
3. W. J. WALTERS, Oil Gasification, *Gas Times*, **83,** 81, 158 (1955)
4. D. J. KING, Carburetted water gas from heavy fuel oil, *Coke and Gas*, p. 201 (1954)
5. J. H. DYDE, New developments in the production of town gas, *The Gas World*, 143, 580 (1956)
6. R. B. B. COX and H. JOHNSON, Fuel oil gasifiers for industrial use, *Proc. Conf. on Major Developments in Liquid Fuel Firing*, Inst. Fuel (1959)
7. E. R. WARD, Oil and its derivatives for town gas manufacture, *Proc. Conf. Major Developments in Liquid Fuel Firing*, Inst. Fuel (1959)
8. Symposium on the uses of liquefied petroleum gases, *J. Inst. Fuel*, **33,** 111 (1960)
9. C. STOTT, Gasification at the Isle of Grain, *Inst. Gas Engrs. Publication*, No. 568 (1960)

CHAPTER 9

Oil Fired Domestic Heating Appliances

SINCE 1950 there has been a rapid and continued increase in the quantity of fuel oil used for domestic heating. Whereas oil fired central heating was previously confined to larger houses, in recent years a demand has arisen for oil heating to varying standards of comfort in houses of all sizes. The equipment used varies from portable kerosine convector and radiant heaters, through space heaters and ducted hot air systems, to boiler installations supplying full central heating. The growing popularity of oil is due mainly to the running costs for an oil fired installation becoming more competitive with solid fuel installations and the advantages over solid fuel of convenience, cleanliness and ease of control.

Oil Burners Used in Domestic Heating Appliances

Vaporizing Burners

Vaporizing burners are normally used in portable air heaters and small water heating or central heating boilers. The oil must be vaporized from a film and sufficient combustion air must be mixed with the vapours. The film of oil may be provided either by means of a wick, a carburettor pot, or a centrifugal spinner, and whereas kerosine is essential for most types of vaporizing burners, a 35 second gas oil (domestic fuel oil) can be used for the pot type of burner. Semi-automatic vaporizer burners using the high flame/low flame or the high flame/pilot flame principle have been popular in the U.K., but the trend is towards the fully automatic self-igniting vaporizers, since the latter consume no fuel when the boiler thermostat calls for no heat release in the boiler.

TABLE 38

Typical Properties of Kerosine and Domestic Fuel Oil

	Kerosines		Domestic fuel oil
	Premium	Regular	
Specific gravity, 60/60°F	0·80	0·82	0·84
Flash point, closed, °F, min	110	110	150
C.V., gross, B.T.U./lb	20,000	19,900	19,600
Viscosity at 60°F, cs.	2·0	2·1	5·5
Conradson carbon residue, 10% bottoms, % wt	Negligible	Negligible	0·05
Sulphur, % wt	0·05	0·05	0·8
Char value mgm/kg	7	15	
Smoke point, mm	40	25	

Annular wick-type burner.—In this type of burner (Fig. 39) the oil vaporizes off the surface of the annular wick. It flows up the wick by capillary action and is vaporized at the surface of the wick by heat transferred from the flame. The rate of burning depends on a number of factors, including the amount of exposed wick, the height of the top of the wick above the level of the oil in the reservoir, the size and nature of the wick, and the viscosity and surface tension of the oil. The fuel in the reservoir becomes slightly warmer after the heater has been lighted and the flame size usually tends to increase as a result. The annular wick type of burner is popular in flueless convector heaters and premium grade kerosine is required in order to avoid smoky combustion and the deposition of excessive char. Combustion is complete and the products are discharged into the room in which the heater is placed, the efficiency being 100% on the basis of the net calorific value of the fuel.

Short drum vaporizing burner.—This type of burner is quiet in operation and gives a blue soot-free flame. It creates its own draught and small units are fitted in flueless appliances. Larger units have a chimney fitted for the removal of the products of combustion and to provide increased draught where this is necessary. The burner (Fig. 40) consists of one or more pairs of concentric perforated drums placed above interconnecting grooves which are fitted with kindler wicks to initiate combustion. The vapours are ignited with a torch and, as the drums warm up, the

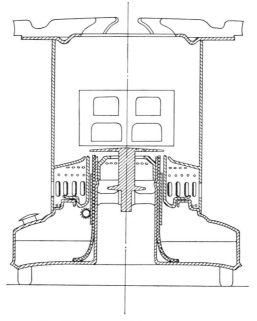

Fig. 39. Annular-wick portable stove.

increased quantity of vapour produced rises to a level in the annular space where sufficient air for complete combustion has been induced. The characteristic blue flame results. The annular space between pairs of drums is covered by a metal or refractory lid in order to direct all the air induced by the chimney effect of the hot burner to enter the gas annuli and mix with the vaporized oil. These covers also provide extra radiating surface. Although kerosine is the recommended fuel, blue flame appliances are less affected by the hydrocarbon type and smoke point of the kerosine used than are wick type burners. Careful installation and levelling are necessary, and the fuel feed must be arranged so that the oil

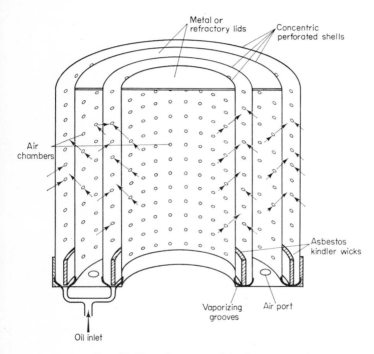

FIG. 40. Short drum vaporizing burner.

is always at the required level in the grooves. A bird-bath or a float carburettor type of feed is used, the burning rate being controlled by a metering valve or on the high flame/low flame principle using a thermostat. The period on high flame is controlled by the boiler thermostat. Instead of operating on low flame, ignition can be provided by means of small pilot wicks, which are arranged to touch the kindler wicks and consume as little as 1/20 pint per hour. Automatic safety control to safeguard against flame failure or flooding can be provided, and is usually incorporated in the controller. Automatic electric ignition has been successfully applied to a special design of short drum burner. The larger size of burner (40,000 B.T.U./hr) has two pairs of concentric shells, and where a higher output is required, two burners, each with its own fuel control, are provided.

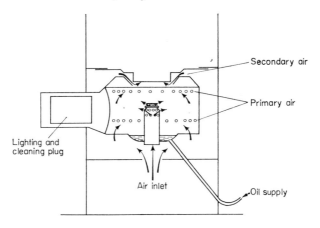

Fig. 41. Natural draught pot type burner.

Pot burners.—The pot, or carburettor burner (Fig. 41), comprises an oil reservoir and vaporizing chamber. In the type shown, air is introduced through a central pillar and progressively through holes in the pot, the spacing of the holes being critical in order that the mixture is too rich to burn on or near the oil surface. The heat radiated from the flame and conducted through the metal to the oil pool is sufficient to vaporize the fuel at the required rate, the vapours being mixed with the primary air and further mixed with secondary air to burn at the top of the pot.

The rate of burning depends on the amount of oil vaporized, which in turn depends on the temperature and size of the pot, the distance of the flame from the oil reservoir, and the amount and velocity of air drawn into the burner. Draught is therefore a further critical feature of these burners and the installation of an automatic draught regulator is essential when natural draught is to be used. The rate of oil supply must be accurately controlled, and a safety device is necessary in order to prevent the oil in the pot from rising above a safe level. As with the drum type of burner, careful levelling of the pot is essential.

A forced draught pot burner is in principle similar to the natural draught type, but the air is supplied from an electrically driven fan (which consumes about 40 W) thus enabling high rates of vaporization to be achieved with good control of air. The air supply is not entirely independent of chimney draught and automatic draught regulators should be installed. These burners generally operate on the high flame/low flame principle, but full air flow is usually continued during low flame operation and poor efficiency results throughout this condition. In order to obtain high overall combustion efficiencies it is necessary therefore to select a burner which operates for maximum periods on the high flame setting. A recently developed fan-assisted pot burner operates on the high flame/off principle, the oil being ignited electrically and the fan running only for a short purge period after the oil supply has been automatically cut off. Such a burner should be more efficient when working on less than full capacity than is the type of fan-assisted burner in which only the oil supply is reduced on low load.

Although gas oil can be used with the pot burner, kerosine is the preferred fuel since less carbon deposits are formed and the burner would thus require less cleaning. Carbon formation is less in the fan-assisted burner than in the natural draught type, but the former suffers from the disadvantage of flame roar. The maximum output from a fan-assisted pot burner is about 80,000 B.T.U./hr.

Wall flame rotary vaporizing burners.—This design of forced draught burner (Fig. 42) has been in use in the U.S.A. for many years but has only recently been introduced into the U.K. A kerosine fuel is uniformly distributed on to a stainless steel wall by means of a centrifugal spinner, the wall being heated by radiation from the flame, and vaporization occurring at the wall and from the coarsely atomized spray produced by the distributor. The fan and fuel distributor are driven by the same motor. The

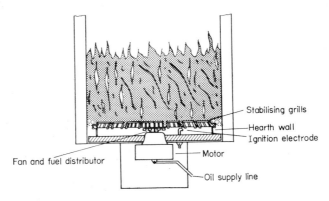

Fig. 42. Wall flame rotary vaporizing burner.

air is directed towards the vaporizing wall where it is deflected upwards, mixes with the vapour and combustion takes place from the stabilizing grills placed above the wall. The mixing of air and vapour is extremely good and high combustion efficiency results. The low heat capacity of the vaporizing wall, together with the small amount of fuel distributed per unit area of wall, make it possible to ignite the vapours by means of an electric spark formed between a single electrode and the wall. Since the air and fuel supply are both cut off by the thermostat when no heat is required, the efficiency of this type of burner does not decrease as much as with the usual type of pot burner if the loading is appreciably below the burner rating. The burner can be used for the 40,000 to 80,000 B.T.U./hr range, has a low power requirement, and is quiet in operation.

Atomizing Burners

The basic principles of the operation of atomizing burners are dealt with in Chapter 6, and only their application to domestic appliances will now be discussed.

Pressure jet burners.—Pressure jet burners are most satisfactory for installation in the larger domestic appliances, but when the required output is below about 60,000 B.T.U./hr the pressure jet is severely limited for the following reasons. The nozzle of such a small burner, delivering approximately 0·6 gallons per hour, would be less than 0·01 inch in diameter, and very accurate machining is necessary if satisfactory operation is to be obtained. Furthermore, such small nozzles are easily choked and have to be used in conjunction with very fine filters, which themselves are easily choked.

Pressure jet burners are simple and relatively cheap. They are comparatively quiet in operation although sometimes noisier than vaporizing burners. With the improved types of combustion heads developed in the last ten years, smokeless combustion can be achieved with as little as 15% excess air. Because the output variation from a pressure jet is small they are operated on the on/off principle. Automatic shut-off valves and delayed opening nozzle valves are fitted to prevent inferior combustion conditions and hence fouling of the boiler surfaces when the fuel is admitted to or shut off from the burner. The most suitable fuel for use in domestic pressure jet burners is gas oil, since if kerosine is used its poor lubrication properties might lead to excessive wear in the fuel pump.

It must be remembered that, whereas a pressure jet or other type of automatic intermittently operating burner is self igniting and uses no fuel in its off periods, boilers suffer from a loss of efficiency due to intermittent operation and the working of the appliance below its full rating. This is illustrated by the curve of Gollin, reproduced in Fig. 43. During the off-period of intermittent operation there is an external loss of heat from the boiler which is increased due to the cold air passing through it. The curve shows the effect of this loss together with the loss in efficiency encountered (even in continuous operation) when a boiler is worked at loadings which are appreciably below their full rating. There remains a great deal of controversy on this subject.

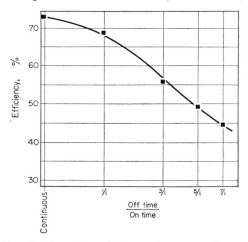

FIG. 43. The efficiency of intermittent operation.

Air atomizing burners.—This type of blast burner, although capable of operating satisfactorily at half the output of the smallest pressure jet and on slightly heavier fuel, is more expensive and noisier. Thus they are superior to pressure jets only in the smaller type of appliance. Air atomizing burners operate with air at a pressure of about 5 psig, about 5% of the combustion air being supplied by a compressor for atomizing the fuel, the remainder being supplied by a low pressure fan or by natural draught. The former method is preferred since the use of natural draught results in a variable air supply to the burner.

Burner Controls

Constant Level and Metering Controls for Vaporizing Burners

A common type of control is shown in Fig. 44. The inlet needle valve is balanced by a float and thus maintains a constant level in the controller. A spring-loaded safety trip device is incorporated so that if the oil rises on failure of the float controlled needle valve, the trip mechanism operates and applies the spring pressure to close this valve. The trip device, which is also operated by the

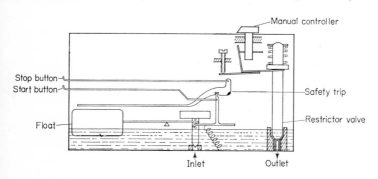

Fig. 44. Constant level control for vaporizing burner.

"stop" button on the controller and is reset by means of the "start" button, also acts as a flame failure device. Should the flame be extinguished the oil level will rise in the burner and in the controller until the trip mechanism shuts the valve. The oil is supplied from the tank via a filter and is metered from the constant level controller by means of a restrictor valve in the outlet. The quantity of oil fed to the burner can be regulated manually or automatically from a thermostat. The automatic operation can be electrical, using the bending of a bimetallic strip to raise or lower the restrictor, or mechanical, by means of a metallic bellows connected via a capillary to a temperature sensitive phial. The electrically operated control provides high flame and low flame settings, whereas with the mechanical type the restrictor position can be made to modulate. Fan assisted vaporizing burners should have a solenoid valve between the burner and the control, so that the oil supply is isolated in the event of an electricity failure and thus avoiding the poor combustion conditions which would result due to an inadequate air supply.

Flame Failure Devices

For vaporizing burners, the constant level controller shown in Fig. 44 acts as a flame failure device. For atomizing burners, the flame failure device may be based on the temperature of the products of combustion (fluestat) or alternatively by means of a photo-electric cell or thermopile which is positioned to receive radiation from the flame. Whereas the types which are based on flue gas temperature are somewhat sluggish in operation, those based on flame radiation are very rapid and are preferred in large appliances having high oil consumption rates.

Oil Burning Appliances Used in Domestic Heating Installations

Space Heaters

Approximately 11 million portable kerosine heaters of either convector or radiant type are now in use in the U.K., mainly for supplementing other sources of heat and for use in rooms used for short periods. These appliances have heat outputs of up to 10,000 B.T.U./hr and are flueless, the waste gases exhausting into the room. The thermal efficiency based on the net calorific value is 100%, but more ventilation than normal is generally desirable.

The small convectors have annular wick burners, whereas the radiant heaters use small short drum vaporizer burners to heat a hemispherical wire gauze (Fig. 45). The gauze becomes red hot and provides a proportion of the total heat output as radiant heat, a metal reflector being placed behind the top of the burner.

This drip feed radiant type of heater was the subject of the Oil Burners (Standards) Act of 1960, which makes provision for minimum standards of efficiency and safety for flueless kerosine space heaters. It had been shown that radiant heaters could give rise to serious fires if subjected to strong draughts. In this condition downdraught causes the burning to be transferred from the top section of the shells to the base. Flames had been found to spread to the fuel tank, resulting in the spillage and ignition of kerosine. Recommendations[7] to avoid such mishaps include the fitting of an extra cylindrical shell sealed to the base plate which should not be perforated within the cylinder.

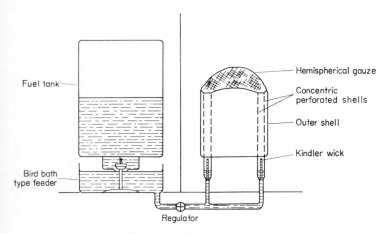

FIG. 45. Radiant oil heater.

A number of the larger type of flued free-standing space heaters are marketed, these having a heat output of 15,000 to 75,000 B.T.U./hr and using kerosine or gas oil as the fuel. These appliances which generally use a pot type short drum or sometimes pressure jet burner, can be used for the heating of open plan houses, workshops, and the like, but have a limited application depending on the planning and lay-out of the building. Forced convection units are available for central heating by means of duct circulation, but owing to the complicated pattern of movement of warm air within a house, such systems can be less economical to operate than central heating by means of hot water circulation. However, they are becoming more popular due to their lower initial cost and their ability to provide rapid heating up of the air.

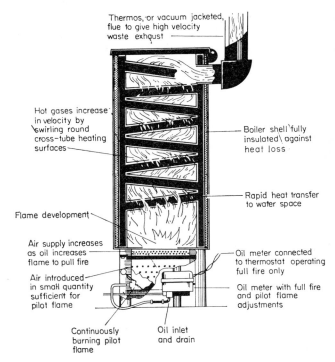

Fig. 46. Boiler–burner unit employing pot type burner.

Hot Water Boilers

Although the early use of oil in domestic heating boilers was in converted solid fuel appliances, the demand for oil fired equipment has progressed so quickly that there is now available a wide range of boiler burner units which have been designed specifically for oil firing. These boilers usually have higher efficiencies than conversions, since they have a higher ratio of secondary convection heating surface to primary surface than is provided in solid fuel boilers. Thermostatic control is incorporated, either by intermittent operation with automatic ignition, by high flame/low flame operation, or by the use of a high flame/pilot system.

For boilers up to 60,000 B.T.U./hr rating vaporizer burners are used, and for larger boilers the most common burner is the pressure jet. It is important to select a boiler which operates for maximum periods at the full rating, since with all types of burner there are disadvantages and a reduction in efficiency at low loads.

The smaller sizes of boilers use fan assisted or natural draught pot type burners, or the short drum burner. The pot burner gives little radiant heat and therefore requires a relatively large proportion of convection surface in order to achieve high efficiencies. It is also dependent upon the provision of good chimney draught. The short drum burner is generally preferred, since it is not so sensitive to chimney draught and, owing to its high proportion of radiant heat needs less convection surface and results in a more compact boiler. Examples of both types are shown in Figs. 46 and 47.

Complete boiler-burner units designed for pressure jet burners provide the necessary heating surface by means of vertical or horizontal gas passages, or are essentially sectional boilers with adequate secondary surface. The efficiency of such units is generally high, 75 to 80% being possible at full maker's rating.

The Storage of Domestic Fuel Oil

A domestic storage tank should be made of ungalvanized mild steel, internally braced and well supported. The capacity should be determined by the maximum rate of oil usage, consideration being given to the price of the fuel when supplied in different quantities. The cheapest price is for quantities of 500 gallons or more, and the smallest delivery is normally 100 gallons.

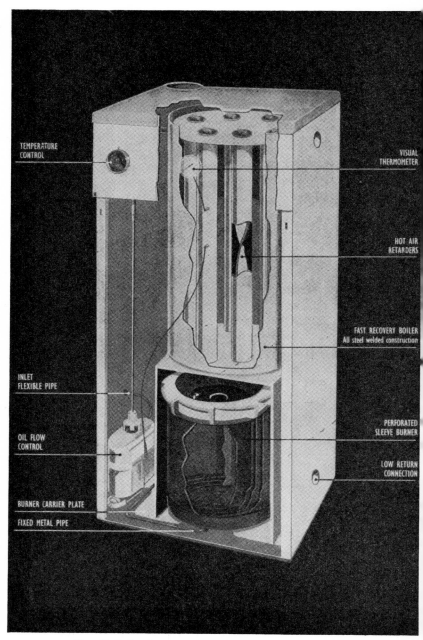

Fig. 47. Boiler–burner unit employing short-drum burner.

The tank (Fig. 48) should be fitted with a filling line, a drain valve, a draw-off point, and some means of indicating the quantity of oil in the tank. The fill pipe inlet should be fitted with a screwed

Fig. 48. Small domestic fuel oil tank.

cap, and the vent pipe should be the same size as the fill pipe with its open end turned down and covered with a wire mesh. Where the supply vehicle is fitted with a reeled hose and trigger nozzle, separate fill and vent pipes are necessary only when the fill pipe is extended from the top of the tank.

The drain valve should be fitted at the lowest point on the tank in order to enable accumulated water and sludge to be drawn off periodically.

The draw off valve should be placed above the bottom of the tank to allow space for the water and sludge to collect and to ensure that none of this material is drawn into the burner.

Several types of gauges are available for indicating the contents of the tank, the most common incorporating some type of float. Owing to the possibility of breakage, gauge glasses should not be used unless adequate safety shut-off valves are installed. Atomizing and vaporizer burners have fine filters supplied, but it is recommended that a slightly coarser filter can be fitted in the supply line to the burner. It is good practice to fit a fire valve in the line, so that in the event of a fire originating in the house, the oil supply is automatically shut off. Care is necessary to ensure oil-tight joints in pipework from the tank to the burner.

Central Heating

Central heating by means of ducted hot air is not very popular in the U.K. and most installations employ hot water in either natural or forced circulation radiator circuits.

The development of the small pipe forced circulation system[4] has done much to popularize central heating for the smaller type of dwellings. A small, quiet, glandless water pump is used to circulate the hot water through $\frac{1}{2}$-inch pipes, and automatic room temperature control is achieved best by means of a mixing valve which is actuated by outside air temperature and which determines the temperature of the water flowing to the radiators.

Whatever the type of central heating system, it must be capable of supplying heat to maintain the various parts of a house at a constant selected temperature during the coldest weather (taken in the U.K. as 30°F). The heat loss for each room through walls, floors, windows and roof can be calculated, together with the air ventilation loss, and a sound basis for determining the size of the heating appliance is thus obtained, due allowance being made for domestic hot water requirements. Many heating engineers add too many safety factors when determining the size of appliances, and oversizing is common. Care should be taken to ensure that the appliance is of adequate but not ample size for the duty, since oversizing adversely affects the initial cost and operating efficiency.

References

1. H. ROPER, Oil fired domestic heating, *Heating*, Sept. 1958 to Feb. 1959
2. K. H. SAMBROOK and N. M. LAWRENCE, Liquid fuels for domestic heating: A ten year review. *Major Developments in Liquid Fuel Firing* 1948–59, Institute of Fuel
3. A. B. PRITCHARD, Oil fired central heating, *The Industrial Heating Engineer*, June 1956 to May 1957
4. D. V. BROOK, Small pipe central heating; design, installation and cost. *The Architects Journal*, Oct. 17th and 24th, 1957
5. G. J. GOLLIN, Oil fired appliances for domestic heating. *Special Study of Domestic Heating in the U.K., Present and Future*, Institute of Fuel, 1956
6. G. J. GOLLIN, Oil fuel and the sectional boiler, *J. Inst. Fuel*, 33, 310 (1960)
7. The Effect of draughts on the burning of portable oil heaters, *Her Majesty's Stationery Office*, 1960

CHAPTER 10

Storage and Handling of Liquid Fuels

LIQUID fuels are stored and handled in systems which consist basically of tanks, pumps and pipelines, the construction of these installations varying with the characteristics of the different fuels.

Storage Tanks

Liquefied petroleum gases are stored at pressures of up to 250 psig to maintain them in the liquid state and the tanks used are either small diameter, heavily constructed long horizontal vessels with convex dished ends, or specially constructed spheres (Fig. 49). Low flash point fuels, such as motor and aviation gasolines are usually stored in tanks fitted with roofs which float on the surface of the oil (Fig. 50). These eliminate any air space above the liquid and reduce fire hazards and evaporation losses when filling or due to "breathing", the latter occurring as a result of changes in vapour pressure with variations in temperature. Blankets of plastic material which float on the oil surface can also be used to reduce evaporation losses when storing volatile fuels. High flash point fuels such as kerosine, diesel oils and fuel oils are stored in tanks which have fixed conical shaped roofs (Fig. 51). Tanks which are used for the storage of the heavier fuel oils are fitted with heating coils in order to maintain the oil at a temperature at which it can be pumped easily.

Pressure vessels are fitted with pressure and vacuum relief valves and also suitable level gauges and sampling devices. Other tanks are fitted with dip hatches, through which the contents can be measured and samples taken, and vents to allow the air to escape

STORAGE AND HANDLING OF LIQUID FUELS 163

when the tanks are being filled. Storage tanks are fitted with manholes in the roof and in the bottom course of plates, these being used for ventilation and access when a tank is opened for maintenance. Tanks are normally sited in oil-tight enclosures to

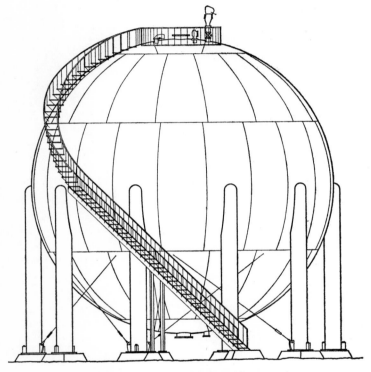

Fig. 49. Elevation of a 15,000-barrel Hortonsphere for pressure storage.

contain any leakages of oil and are protected by fire-fighting equipment such as foam and water lines.

All storage and handling operations are governed by various

164   LIQUID FUELS

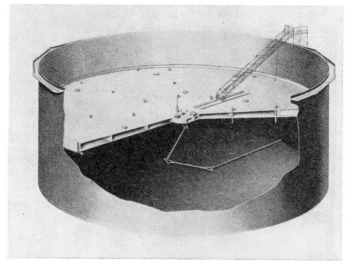

Fig. 50. The Horton double deck floating roof tank.

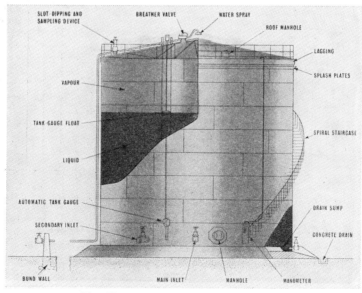

Fig. 51. Vertical tank for bulk oil storage.

safety measures, as outlined in a model code of Safety Practice in the Petroleum Industry.[1]

Liquefied Petroleum Gases

Liquefied petroleum gases are transported from refineries under pressure in special tanks or in cylinders. Bulk deliveries are usually made where the gases are used for industrial applications and the LPG is then offloaded as a liquid into pressure storage vessels at each installation. Liquid is then drawn from these vessels as required, vaporized and piped to the point of utilization. For domestic purposes and in industries where the consumption of LPG is relatively small the gases are supplied in steel cylinders, the capacities of which are usually in the range 10 to 120 lb. These cylinders are fitted with valves and pressure regulators and the LPG can be removed in either the liquid or vapour state.

To prevent the formation of explosive mixtures in storage and handling systems special precautions are taken to remove air prior to the introduction of LPG and also to prevent the introduction of air during use. LPG containers are not completely filled and a vapour space of at least 20% of the total volume of a container is maintained to allow for liquid expansion with increase in temperature. Recommended filling ratios for LPG containers are specified in BS 1736 : 1951.

The minimum pressures required to maintain propane and butane as liquids at different temperatures, i.e. the vapour pressures of these liquids at these temperatures, are shown in Table 39. All containers must be sufficiently strong to withstand the highest pressure likely to be experienced, and recommendations for the materials and methods of construction of steel cylinders for the storage and transport of liquifiable gases are given in BS 401 : 1931. All-welded pipes are recommended for distribution systems and where flexible hoses are used these should be made of special synthetic rubber or other material which is resistant to the solvent action of LPG.

TABLE 39

Vapour Pressure of Propane and Butane

Temperature, °F	Vapour pressure, psi	
	Propane	Butane
32	50	0
60	92	12
100	172	37

Materials Having Flash Points Below 73°F

Motor and aviation gasolines, benzole and ATG* have flash points below 73°F and their storage in the U.K. is, in general, controlled by the Petroleum (Consolidation) Act 1928, and the Petroleum (Mixtures) Order 1929.[2] Under certain circumstances up to 60 gallons of these materials, defined as Petroleum Spirits, may be stored without licence for use in connection with a motor vehicle, motor boat or aircraft, provided they are not offered for resale (Petroleum Spirit (Motor Vehicle etc.) Regulations 1929).

Petroleum Spirit is received at distribution depots from main installations and is subsequently transferred, either by pipeline, or in cans, drums, or road tankers to the consumer, e.g. filling stations and airports. All installations, depots, filling stations and so on must meet various requirements prescribed by local authorities, such as:

(a) All tanks should
   (1) be sufficiently strong to stand severe strains,
   (2) be liquid and vapour tight, except for vents,
   (3) have vents so placed that the vapours will be safely dispersed,
   (4) be sited below ground in the case of filling stations.
(b) No operations involving fires or other sources of ignition shall be carried out near the Petroleum Spirit.

* Aviation turbine gasoline.

STORAGE AND HANDLING OF LIQUID FUELS 167

(c) All pumps and electrical equipment must be flame-proof.
(d) Adequate fire fighting equipment should be installed.
(e) All vessels which have contained Petroleum Spirit and been emptied, but not rendered gas free, should be kept securely closed.

In the case of aviation fuels, filtration prior to storage in aircraft fuel tanks is extremely important to remove fine solid particles and entrained water which could cause abrasion, blockage and corrosion in aircraft engines. There are three main types of filtration equipment using expendable cartridges (e.g. glass fibre), non-expendable cartridges (e.g. ceramics), or expendable filter aids.[4]

Fuel Oils

Fuel oils can be delivered to oil burning installations by road tank wagon, rail tank car, barge or coastal tanker depending on its location. At these installations the handling and storage facilities will depend on the size of the installation and the grade of fuel oil being used. Basically the facilities are the same in all cases, namely bulk storage tanks, suitably sited for taking oil fuel deliveries, and a small service tank sited near to the oil burners. Bulk storage capacity is governed by the availability of oil supplies and the oil firing rate but it is generally recommended that this should be equivalent to three weeks supply at the maximum oil consumption rate. There are no statutory regulations regarding the storage of fuel oils but a "code of practice" is recommended in the British Standard Specification, B.S. 799 : 1953.

All-welded steel storage tanks are normally recommended although concrete tanks, in some cases lined with glass tiles, are also used.[5] All tanks are fitted with filling, vent and offtake lines together with a drain cock, sited at the opposite side to the oil offtake point, to run off water and sludge. These fittings are shown diagrammatically in Fig. 52.

The lighter industrial fuels are usually pumped direct from

storage to one or more burners and require no preheating either in storage or prior to combustion. Heavier, more viscous fuels generally require heating in storage to prevent any solidification and to ensure that they are sufficiently fluid to be pumped with an economical power consumption. Fuel oils are considered to be easily pumpable when their viscosities do not exceed 2000 Redwood 1 seconds although no difficulties are experienced with oils of viscosity as high as 5000 Redwood 1 seconds. In general for

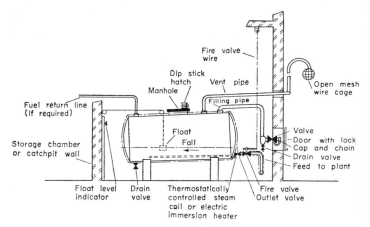

FIG. 52. Diagrammatic arrangement of storage tank for heated grades of petroleum fuel oils

satisfactory atomization the viscosity of a fuel should be about 100 Redwood 1 seconds although for small pressure jet atomizers a lower viscosity is desirable and for rotary cup burners higher viscosities can be employed.

For economic reasons the heavier fuels are usually stored at the temperature at which they will flow to the outlet of the tank and are subsequently heated to their pumping temperatures immediately prior to use. A steam or electric outflow heater inserted through the side of a tank alongside the oil offtake line may be

STORAGE AND HANDLING OF LIQUID FUELS 169

TABLE 40

Fuel Oils—Storage, Pumping and Atomizing Temperatures

Fuel Class	Approx. viscosity Redwood 1 sec at 100°F	Minimum storage temperature, °F	Pumping temp., °F, corresponding to a viscosity of 2000 Redwood 1 sec*	Approx. atomizing temperature (to give a viscosity of 100 Redwood 1 sec) °F*
E	250	45	45	130
F	1000	65	80	190
G	3500	75	110	225

* See Fig. 12.

used for this purpose. Minimum recommended storage temperatures, pumping temperatures and approximate atomizing temperatures for petroleum fuels corresponding to Classes E, F and G in the B.S.I. specifications are given in Table 40.[6]

In the case of the coal tar fuels, the storage and handling of CTF 50 and 100 do not present any special difficulties although crystals may deposit from the latter grade at temperatures below 90°F and it is therefore stored at a higher temperature than the

TABLE 41

Storage of Coal Tar Fuels

Coal tar grade	Minimum storage temperature, °F
CTF 50	40
100	90
200	80
250	140
300	180
400	270

heavier CTF 200. Recommended minimum storage temperatures for coal tar fuels are given in Table 41.[7]

Fuels heavier than CTF 100 are heated before combustion to the correct atomization temperature, i.e. the temperature in degrees Fahrenheit by which each fuel is designated.

Steam or electric tracing of pipelines is essential for the transfer of all coal tar fuels (except CTF 50) and for some residual petroleum fuels in order to maintain the temperature and hence the fluidity of the oil. Small steam pipes or electric heating cables are fitted to the oil lines in order to counterbalance the heat loss via the line lagging to the atmosphere.

Tank Heating

Low pressure steam coils are usually employed for tank heating, though sometimes hot water coils or electric immersion heaters are used, the latter being economical only for small tanks. The amount of heating surface required depends on the quantity of oil in the tank, the atmospheric conditions (wind velocity, temperature and rainfall) and the oil movement in and out of the tank. The theory of heat transfer in tanks and heating coils is adequately dealt with elsewhere.[7, 8, 9] Care should be taken to ensure that the correct data for each tank are available when coil sizes are determined, particular attention being given to likely wind velocities, and the fact that the coil should be capable of maintaining the temperature of the contents of the tank in the most severe conditions. Oversizing of tank heating coils is of little consequence, since temperature control of the contents of the tank is readily applied to the steam supply. Such control is desirable on economic grounds alone, since the heat loss increases rapidly with increase in temperature, and maintaining the oil at its minimum storage temperature is essential if excessive steam consumptions are to be avoided.

The necessity of lagging storage tanks must be assessed in each individual case, since the economics of installing and maintaining the lagging must be weighed against the heat saved. The higher the storage temperature of the oil the greater the saving due to lagging, and in general it becomes economic for tanks in which oil is stored at 100°F and above.

Heating coils should be situated as low as possible in a storage tank, preferably below the level of the oil offtake, with a continuous fall to enable condensate to drain to a trap. An air bleed should also be provided to enable air to be removed from the system.

Oil Preheaters

It is desirable from the standpoint of heat economy to heat the oil to the minimum temperature at which it can be easily pumped and to heat to a higher temperature immediately before atomization. The recommended temperatures for efficient atomization and combustion are given in Tables 40 and 41 but it should be noted that these temperatures are to be used only as a guide. The optimum degree of preheat should be determined by plant trials and therefore preheaters should be capable of achieving somewhat higher temperatures than given in the above tables. Steam and electric preheaters are used, the former being cheaper and more easily maintained. Heaters should be designed such that the heating surface does not operate at too high a temperature, the maximum loading of electric heating elements being limited to 10 W/in^2 and where steam is used for heating it should preferably be dry, saturated at not more than 50 psig.[5] The steam or electric power supply to the heaters is controlled to maintain a constant oil outlet temperature, and thus minimize the viscosity variation. Direct viscosity controllers are available, but these are expensive and not justified unless fuels of widely different viscosities are used in the installation.

Ring Main Systems

Whereas a single pipe system of distribution can be employed for one or more burners handling a light fuel oil, a ring main system is preferred for medium or heavy grades of oil as a means of obtaining the minimum pressure and temperature difference between burners. A flow diagram for a typical ring main system with duplicate pumping, heating, and filtering units is shown in Fig. 53. The oil is filtered on the suction and discharge sides of the pump, the coarse filter on the suction being to protect the pump, and the fine filter on the discharge to protect burner orifices and control valves. Duplex or self-cleaning filters are necessary to ensure efficient filtration with continuous plant operation. A vent pipe is fitted at the highest point in the system

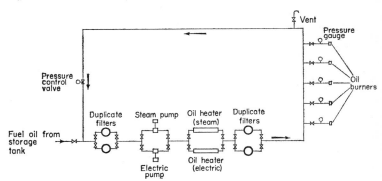

Fig. 53. Diagram of a typical ring main system.

to permit air to be removed when the system is filled and also to permit the main to be completely drained when necessary.

Positive displacement or screw type pumps are the most satisfactory and the rate of oil circulation should be two to three times the quantity being burned, the excess being returned to the suction of the pump. The pressure at the burners is controlled by a pressure regulating valve and pipe sizes should be chosen to give a small friction loss and thus a reasonably uniform pressure at each burner. The circulation rate normally corresponds to an oil velocity of 3 to 5 feet per second which is in the streamline flow range, so that the Poiseuille equation can be used for calculating the pressure drop.[8, 9]

Two pressure ranges are employed, namely low pressure (not exceeding 50 psig) for blast or rotary atomizers and high pressure (up to 700 psig) for pressure jet burners. It is more important to minimize the friction loss in low pressure systems, since variation in pressure has a proportionately larger effect than in high pressure ring mains.

Figure 53 shows a hot oil ring main, in which the oil being circulated is maintained at the temperature corresponding to the viscosity required for atomization. Alternatively, the oil can be circulated at the temperature corresponding to the pumping viscosity and heated to atomizing viscosity by branch line heaters placed near individual burners. The latter method minimizes heat losses but requires larger pipe sizes and more heaters, so that hot oil circulation is generally cheaper to install but line lagging is more important, particularly for the heavier grades of fuel.

References

1. *The Institute of Petroleum Model Code of Safety Practice in the Petroleum Industry*, Institute of Petroleum, London
2. H. E. WATTS, *Storage of Petroleum Spirit*, Charles Griffin (1951)
3. *Storage of Petroleum Spirit and Petroleum Mixtures*, Fire Protection Association, London (1960)
4. J. M. TYGRET and ROBERT ULRICH, *Oil and Gas Journal*, **59,** No. 37, 106, 11th September (1961)
5. T. CHIPPINDALE, The Journal of The Institution of Heating and Ventilating Engineers, April (1957)
6. Proceedings of the Conference on the Major Developments in Liquid Fuels ; p. A-3, Institute of Fuel, 1959
7. *Coal Tar Fuels*, Ch. 4, Association of Tar Distillers, 1960
8. D. Q. KERN, *Process Heat Transfer*, McGraw-Hill (1950)
9. W. H. MCADAMS, *Heat Transmission*, McGraw-Hill (1954)

Index

Additives
 fuel oil, 91
 gasolines, 59, 60, 63
 turbine fuels, 116
Air
 combustion, 87
 excess, 86, 91, 101
 preheated, 81
Air atomizing burners, 81, 99
 in domestic appliances, 152
Alcohols, 53, 64
Alkylation, 9
Anti-knock, 9, 50
 alcohols, 64
 aviation gasolines, 60
 benzole, 31
 motor gasolines, 9, 53
 tests, 52, 60
 vaporizing oil, 63
Aromatics, 29, 113
 removal, 18
Ash, 47, 72, 94, 106, 109, 115
Atomization, 76
Aviation gasolines, 59, 112, 162, 166

Benzole, 25, 29, 53, 166
Blast furnaces, 106
Boilers
 hot water, 156
 packaged, 98
 shell type, 98
 water tube, 96
B.S.I. Specifications
 diesel fuels, 73
 fuel oils, 7, 167
Bubble cap tray, 3
Burners, 76
 blast atomizing, 81

Burners (*Contd.*)
 pressure jets, 78
 rotary, 80
 vaporizing, 144
Burning tests, 45
Butane, 6, 140

Calorific value
 coal tar fuels, 49
 petroleum fuels, 39, 49, 114
Carbon residue, 12, 43, 72, 127, 132
Carburetted water gas, 123
Carburettor icing, 57
Catalytic processes
 cracking, 10, 131
 desulphurization, 14, 20
 gasification, 131
 reforming, 12
Central heating, 160
Cetane number, 70
CFR Engine Tests, 58, 60
Char value, 45
Cloud point, 42
Coal
 carbonization, 24
 gasification, 32
 hydrogenation, 31
Coal tar, 26
 fuels, 27, 73, 169
 sulphur, 47, 103, 106
Combustion, 76
 control, 103
 compression ignition engines, 68
 efficient, 86
 gas turbines, 113
 partial, 122, 134
 spark ignition engines, 50
Compression ignition engines, 65

Conversion processes, 9
Copper chloride treatment, 16
Corrosion
  additives, 91
  low temperature, 89
  stack, 92
Cracking, 5, 8, 10
  catalytic, 10, 131
  thermal, 10, 122
Crude oil, 1

Dayton process, 135
Density, 38
DERV, 75
Diesel engines, 69
Diesel fuels, 69
Diesel index, 71
Distillation, 2
Domestic fuel oil, 145

Emissivity, 28, 95, 103
Engine knock, 50, 60, 62, 68
Engine testing, 58, 60, 70

Fire point, 41
Fischer–Tropsch synthesis, 26, 32, 74
Flames, 76
Flame failure devices, 154
Flash point, 41, 115, 166
Floating roof tanks, 162
Fuel oils, 7, 27
  storage, 167
Furnaces, 99
  temperature control, 101
  pressure control, 102

Gas turbines, 108
Gas manufacture, 118
Gasification processes
  carburetted water gas, 123
  Dayton, 135
  Hall, 128

Gasification processes (*Contd.*)
  Jones, 126
  O.C.C.R., 102, 138
  Onia–Gegi, 131, 134
  Segas, 131
  Shell, 136
  Urquhart, 138
Gasoline
  aviation, 59, 113
  benzole in, 30
  cracked, 12
  from coal, 32
  from shale, 25
  motor, 53
  reformates, 14
  storage, 162, 166
Girbotol process, 16
Glass furnaces, 105

Hall process, 128
High pressure burners, 81, 85, 106
Hydrogen
  Platformer, 13
  processes, 20
Hydrogen sulphide removal, 16, 34
Hydrogenation
  of coal, 31
  in gasification, 123
Hydrofining, 21, 73

Jones process, 126

Kerosine
  aviation fuel, 112
  heaters, 145
  properties, 145
  SO_2 treatment, 18
Knock, 50, 68

Lagging of storage tanks, 170
Liquefied Petroleum Gases (LPG), 6, 119, 139
  use in gas manufacture, 141

INDEX

Liquefied Petroleum Gases (LPG) (*Contd.*)
  installation, 142
  storage, 162
Low pressure burners, 81, 83, 98, 101
Lurgi gasification process, 33

Medium pressure burners, 81, 84, 98, 106
Mercaptans, 16
Motor gasolines
  additives, 59
  composition, 9, 53
  engine testing, 58
  volatility, 54

National Benzole Association
  specification for motor benzole, 30

O.C.C.R. process, 102, 138
Octane number, 9, 10, 14, 31, 52, 58, 60, 62, 64
Oil shales, 23
Onia–Gegi process, 131, 134
Open hearth furnace, 103

Performance number, 62
Platforming, 14
Polymerization, 9
Pot-type burner, 148
Pour point, 42
Pressure control in furnaces, 102
Pressure jet burners, 78, 97, 106
  in domestic appliances, 151
Pressure vessels, 162
Producer gas, 120
Propane, 6, 140

Rectisol process, 34
Refining of petroleum, 1
Reforming, 9, 12
Ring main systems, 171
Rotary burners, 80, 98

Safety in petroleum industry, 165
SASOL, 33
Segas process, 131
Shale oil, 23
Shell process, 136
Short drum vaporizing burner, 145
Smoke, 77
  measurement, 87
  Smoke point, 45
Smut formation, 92
Sodium, 47, 116
Solutizer process, 16
Space heaters, 154
Specific gravity
  fuel oils, 38
  gases, 120
Stability of fuel oils, 48
Stack solids, 88
Steam atomizing burners, 81, 97, 103
Steam heating coils, 170
Storage
  domestic fuel oil, 157
  tanks, 162, 167
  temperatures, 169
Spark ignition engines, 50
Sulphur
  fuel oils, 89, 103, 106, 108
  diesel fuels, 73
  gas manufacture, 125, 131
  gasolines, 59
  removal, 16, 20
  significance, 46
Sulphur dioxide process, 18
Sulphur trioxide, 89
Surface ignition, 51
Sweetening processes, 16
Synthesis gas, 26, 32, 121

Tanks
  heating, 168, 170
  storage, 162, 167
Temperature control in furnaces, 101
Tetra-ethyl lead (TEL), 53, 59, 63
Tetra-methyl lead (TML), 59
Thermal cracking, 8, 10, 122
Thermal reforming, 9

Town gas, 120
Tractor vaporizing oil, 19, 63
Turndown ratio, 79

Urquhart process, 138

Vanadium, 47, 97, 106, 116
Vaporizing burners, 144
   controls, 153
Vapour lock, 55
Vapour pressure, 55, 57, 63
Vis-breaking, 8

Viscosity, 39
   atomization, 76, 79
   coal tar fuels, 27
   pumping, 39
Viscosity index, 40
Volatility, 54

Wall flame rotary vaporizing
   burner, 150
Warm-up performance, 57
Water in fuel oils, 48, 113
Wick-type burner, 145
Wobbe index, 120